Svarog

Builder Paul Anderson.

Dedicated to those who have a go.

Tarmac's Broken Dreams.

"What you lookin' for, joss stick holders?" The voice drifted into my awareness from the left. It was none other than Paul wandering down an isle in Tesco supermarket. He has a quick wit that inevitably twangs the chuckle muscle. We usually talk of bikes, and creating them when we meet. One time I had said I would like to build something electric, his immediate remark was, "Maybe a shaver of some sort." That's just the way he is. Okay, so I have hippy tendencies and a beard, I can't deny it. The lights in my office are all solar powered and I've got a compost bin, but why not? It's not lifestyle yogurt weaving.

Anyway, the conversation turned to the Land Speed Record and why Paul *wasn't* going to go for it. However, I will tell you what was said later. First I want to tell you about Svarog, as it's a story that *needs* to be told. I started the book way before the planned attempt, and have left it as originally written. I feel it gives the best representation of the circumstances that led to the outcome. As you may no doubt know, the plans we make are just a guide for life, we don't have real control over our future, just the impression that we do. As has been remarked by minds older and wiser than ours, "Destiny is in the lap of the gods." "What will be, will be." And that all time favourite, "Many a mikle makes a mukle."

Anyone who has ever ridden a motorbike knows the feeling of exhilaration from acceleration. Going just that little bit faster, getting just that ounce more enjoyment is the goal of many of us. This book brings you an account of an *attempt* to be the fastest man and his team of dedicated petrol heads. Learn what it takes to endeavour to be the fastest and maybe be inspired to make your own mark in the world of two wheeled record holders. Also, be aware of the pitfalls. Our future is *probably* not mapped, therein lay the paradox. But if it is, who is the navigator and has he got the map the right way up?

This story started with an end in mind. As stories go it is a good one but it has that all important unexpected twist. As anyone who has blatted down the quarter mile will tell you, That ten seconds takes a year or two to build up to.

He who is brave is free. Lucius Annaeus Seneca.

Introduction.

Our decision to walk creates the path ahead. Paulo Coelho.

Travel broadens the mind, and this whole 'thing' was quite a journey. This is a story about life, the possibilities and impossibilities, the highs and lows, the reasons why *and* why not. Thin tendrils of chance wind and worm their way out into the world, and many unexpected occurrences would befall the team. At the centre of the Universe is man, everything else happens outside. Then there's tarmac.

Not only are we reaching out into deep space with telescopes, but scientists are also finding ways into the human mind and genes that would freak our forbears. Technology has mushroomed into more than could have been foretold by anyone from thirty years ago. It seems, even science fiction is having a hard time keeping up with what is available to us. We live in times where every bit of Arcane knowledge you can imagine is available. What was once hidden is now accessible to all of us.

Think of all the kings of history and each of us in the Western world has more than any of them. If you were able to prove to a Pharaoh that you could talk to someone from the other side of the world. Demonstrate that you could get on a flying metal bird and see them in a matter of hours and on the way you could download knowledge that would enable you to build a metal horse, that Pharaoh would not raise himself from the ground at your feet. Yet we take mobile phones, TV, ready meals and toilet paper for granted. We complain about the speed of our lap top or how long it takes to microwave a pot noodle. Even the last King of England (George VI) could not have imagined *anything* that we have in our living rooms today. These things are magical to the men of old.

Magic in the form of an internal combustion engine is the subject of this book. I always thought that if I were to write a book, it would have to be about custom bikes, and choppers in particular. Being only twelve in 1969 there was no way to watch an X rated film like Easy Rider, but I did manage to get hold of issue 242 of 'Mad' comic which did a spoof. If I went to WH Smiths I would always check out the poster of the film. Nevertheless, the course was lit and as soon as I got my first bike it had a smattering of rattle can paint to make it mine. That credo has remained part of my life ever since.

I guess that Svarog *could* be catalogued as a chopper. Rigid rear end, all the crap taken off, altered frame geometry, great paint job, go faster bag of bits all bolted onto the right places. The list goes on, plus it was built and ridden by the owner. It seems to tick all the right boxes so far.

Chapter 1
History part 1

Winners never quit and quitters never win.
Vince Lombardi

In compiling this section of the book, some interesting little snippets arose. A long held belief amongst many is that Herr Benz invented the first car in 1885. But road usage, without the power of muscle, goes way back. In fact it has to be credited to the French of all people. Nicolas Cugnot, a French military engineer, designed a steam powered road going vehicle. With a top speed of six kilometres an hour it was designed as a tricycle and built to transport cannons. The year? 1769.

And, if you are interested, two years later he built another, lighter car that he ran into a wall. The first motorised accident had been recorded. Well done that man!

That aside the first land speed record, although unsanctioned at the time, is now being credited to Rivaz.

Date: 1813

MPH: 3

Vehicle: Isaac de Rivaz

Driver: François **Isaac de Rivaz**

Place: Switzerland

Engine Type: Turbine

Information brought to the existing FIA sanctioning body, is to accept Isaac de Rivaz as the first Land Speed Record vehicle, (also inventor of an auto-mobile in 1813) The car was run on a mix of hydrogen and oxygen. It makes one wonder how things would have turned out for the planet if we had followed that route.

However, the widely accepted history starts here:

December 18th 1898. Achères, France. Gaston de Chasseloup-Laubat drove a Jeantaud Duc electric vehicle to the heady speed of 39.24 miles per hour.

That was beaten a month later by Camille Jenatzy on January 17th 1899. Not to be outdone by a Belgian, Frenchman Chasseloup-Laubat got back in his electric car and had another go. I don't know if he had a lighter breakfast or just re-wired the beast, but he got up to 57.65. Jenatzy was not content, and on April 29th 1899 clocked 65.79 miles per hour. Whether it was friendly rivalry or steam curled moustaches and a belligerent squint at dawn, has not been recorded. The thing that was set in stone was the whole ethos of Land Speed Records was well and truly under way.

It was not long before internal combustion engines took to the field to try and wrestle this title from the electrics. 5th of November in Ablis, France, an American by the name of William K. Vanderbilt, (you just know he had 'Old' money in this) drove his Mors Z Paris-Vienne at 76.08 mph. All these early attempts were over one Kilometre as France was the test bed of choice. In fact, except for one shot in Ostend, Belgium it wouldn't be until 1904 that the other parts of the world got a look in. Then on 12th January that year one Henry Ford drove his Ford Arrow on a frozen Lake St. Clair at 91.37 mph. I don't know how that became a *Land* Speed Record as Ice isn't exactly land, but the adjudicators seem to have let that one slide. So to speak.

The Comte de Chasseloup-Laubat set a land speed record of 57.6 mph at Acheres, near Paris on the 4th March 1899 in this formidable vehicle weighing over 3,000lbs or 1,400kg.

That's all very well and good but it's the motorcycle record we are interested in.
Funnily enough the 'British Motorcycle Land Speed Record' seems to be the least documented record of all. Maybe the record itself should go into the Guinness Book for that fact. Internet, library, the portal where the Bermuda Triangle deposits it's catch, that arcane place where *half* your socks go, you name it and I searched. Finding the facts about this most elusive record has been quite a frustrating process. I was beginning to wonder if

there *were* very many British records.

Motorcycle Land Speed Record.

Glenn Curtiss was said to have set, unofficially, the first record in 1903 on his home built
bike. He hit 64 miles per hour with his 1000cc
Hercules V-Twin engined bike.

Bitten by the bug, he had another shot in 1907,
with a 4000cc beast. He topped 136.27 mph and
that stood for another twenty years.

(Note how quadrupling the cubes only doubles
the speed. There may be questions later.)

The first *officially*-sanctioned FIM (**Fédération Internationale de Motocyclisme** or
as we Brits would say *International Motorcycling Federation)* record was not set until
1920.
The FIM didn't come into being until 1904.
1923 saw the first record set in the UK by Bert le-Vac on a 996cc Temple Anzani. With
108.41 mph, he took Brooklands by storm. He beat his own record with a Brough-Superior
in 1924. (118.98mph) But along with Claude F. Temple and Oliver M. Baldwin these early
times have not been sanctioned by the FIM rules so do not formally exist*. It took Joseph
S. Wright on an OEC Temple JAP to take the first official record at 137.23 mph. But that
was in Arpjon, France. The first British Land Speed Record was going to have to wait till
Wright came to Ireland to hit 150.65. But Ireland is independent of Britain since 1949 so it
all gets rather confusing.

*(*It would seem red tape was the order of the day and no matter what one did, there was
always someone with a 'control freak' attitude that wanted a piece of the action. On the
other hand someone might have fibbed and taken a record by stealth. Naughty boys.)*

There was disputation over the 1930 record, when OEC claimed to be fastest, on the basis
of a publicity photo taken, before a Zenith went quicker. It was quite a while before the
controversy died down.

As ever, the technology available in the day would be used to it's greatest advantage. That is what sets this particular endeavour apart. Paul intended to use technology that wouldn't be in production for another five years. Plus it would be aircraft, not internal combustion engine technology as in previous years.

One big question hits people when they first learn of the power plant. Does a V Rod have the 'balls' to do the business? Well, if American Kenny Lyon can set eleven records with a Gold Wing at Bonneville, then any assumption is mere speculation. Totally unfounded.

History Part 2
Svarog The god.

Paul has a great interest in history, and in particular the Byzantine Empire. It was when talking with Andreas, Airbus' top designer, that they came upon the name Svarog as a moniker for the bike. As the conversation unfolded it seemed the perfect name.

The name Svarog goes way back and is purported to be one of the earliest known deities.

The name of Svarog is found in the Hypatian Codex, a compilation rediscovered in 1617 in what is now known as the Ukraine. It contains a Slavic translation of an original Greek manuscript of John Malalas from the 6th century. The complete passage, reconstructed from several manuscripts, translates as follows:

"(Then) began his reign Feosta (Hephaestus), whom the Egyptians called Svarog ... during his rule, from the heavens fell the smith's prongs and weapons were forged for the first time; before that people fought with clubs and stones. Feosta also commanded the women that they should have only a single husband... and that is why Egyptians called him Svarog... After him ruled his son, his name was the Sun, and they called him Dažbog.

In the Greek text, the names of gods are Feosta, Hephaestus and Helios. Hephaestus was the Greek and Vulcan the Roman equivalent of Svarog. (We all know who the Norse god is, we named Thursday after him. No more clues, save to say, his film is very popular amongst the ladies.) Apparently, the Russian translator tried to re-tell the entire story (set in Egypt) by replacing the names with those that were better known to his countrymen. It is uncertain how much resemblance the Greeks gods have to their Slavic counterparts. Or when Disney will make a film about it.

What comes out of this is the representations of scientific principles that these gods portray? As in all lesser deities there are deeper meanings, and Svarog represents the inaugural fire. Much like the Hindu god, Garuda who seems to represent the Big Bang, Svarog represents that inner fire that gives rise to the Sun and all flames on Earth. The creative spark if you will. Imagine a fire that is beyond the flames we see in the third dimension and you are getting close. It may even represent that inner fire that is our life force.

Just tell me if I get too "hippy" for you. Daddio.

Scientific principles run as an energy and ancient man was well aware of these energy patterns. Neuroscientists believe that our earliest relationship with gods portraying the unseen forces of nature would have been around one hundred thousand years ago. It was at that time that the frontal lobes began to develop sufficiently to allow new patterns of thought. Writing and art would have come about from this advancement also.

At first I was thinking of Svarog as a lesser god. Then I remembered someone who had studied all this sort of stuff saying that all the Avatar's, (That is God descended into flesh for the advancement and evolution of mankind. As opposed to someone like Buddha or Jesus who realised their divinity by meditation and insight.) i.e. an incarnation of God, always had the letters 'A' and 'R' in their name. Rama, Krishna, Ishvara etc. Could Svarog have been one of the earlier incarnations ten or twenty thousand years ago?

Don't know, don't care! There is only 'Now' and all that far out jazz.

For now we'll just say he was the inaugural bringer of fire. All in all it's a fitting name for the bike.

svarog-kemerovosud-vestul-
siberieirusia

Inventing the wheel.

The Fire Wheel
Symbol.

(The Byzantine
Racing logo was
painstakingly
designed by
Paul's other half,
Christine.)

Imagine yourself
back twelve thousand years. The world was a very different place and the people were a
disparate race. Commonly known as the Stone Age, these neolithic folks were interested in
one thing only. Survival. In order to do this they had to be in with 'The In Crowd' or life
was short. Although they may have domesticated a few wolves to act as guardians they
really only had themselves to rely upon. There were unknown things lurking *'out there'*
ready to devour them at the earliest opportunity. Not only animals but natural forces.
Mother Nature was even more unpredictable than it is today. Nevertheless, the people
were survivors for one reason only, they were beginning to be intelligent. They would hunt
in co-ordinated groups as physical strength counted for very little against a Sabre Tooth
Tiger or a herd of Yaks. If someone tried to take home the 'shopping' with one of these then
they had very little chance alone. That interaction between clan members inaugurated
greater and greater thought patterns.

As the mind began to develop they saw that energy was at play around them. Not only did
they paint what they wanted to happen on the walls of caves, they also depicted what
worked around them. The forces of the Sun and Moon were very evident. The seasons and
the terrain ruled their world. These thought forms could have no picture to represent them
as the totality of variables was too diversified so a pictogram or representation was the
easiest way to signify what was there. One of the earliest they would use was the Fire
Wheel. It described the rotation of the life giving Sun through the sky. It was made up of

the Hakwnkreuz and the circle. Because the circle was such a basic symbol that defined their lives, they saw it as divine. It represented the journey from birth to death, the seasons and the Sun. It portrayed the forces that ruled over them, namely, God. And this was long before they found that it could be turned into a practical wheel. (Personally I don't think the wheel as a form of transport was invented as much as discovered.) The Hakwnkreuz is thought to be as old as the circle in imagery and symbolism. In Hindi it was written thus; ßvaißtakaã. This design would end up incorporated in many later representations and down through history it has had different names from flyfot to the Gammadion.

How Symbols come about.

Thousands of years ago in the Himalayas There wasn't very much to do except think about things. Especially if you were a dedicated monk. One of the things they would think about was 'thought'. How do we think? What is a thought made up of and how can one thought be more powerful than another. How can thoughts and especially decisions change our life? Their studies led them to see that pictures play an important part in the process. Everyone knows a hippy who has a mandala on their kitchen wall. Don't they?

Mandala's are specifically defined pictures usually made up of squares, triangles and circles. Sometimes with a Hindu, Buddhist or Taoist god in the middle, these shapes and colours are meditated upon to bring about specific results in the one who meditates.

How does it work?

Light, like sound, is an energy. As a tuning fork can set another of the same frequency vibrating, these pictures would set the brain waves off in a similar frequency to the light that was reflected off the mandala. Simple as that. Altered states of awareness through concentration on representations of energy.

Going back in Human history we see that ten thousand years ago men used rudimentary thought. It wasn't as complex as the thought processes we enjoy today. After all, where your next bison was coming from and how could you stay warm were probably two of the main thoughts of the day. The third, which is still uppermost in our minds today is "how do I get laid".

It's unclear when hominids took on language as a form of communication but we know that pictures played an important part in the psyche. We still use that way of thinking today. Pictures are what makes up the majority of our thought patterns. If I ask you what colour your bike is you would get a picture in your head right away. That picture is then translated by the neocortex to the word. It all happens lightening fast. (For most folks.) Although we may see our thoughts running all day long and most of them are words, (usually about something that has not or will not happen,) the basis of those words is a picture of some sort. The whole Universe is built on this fundamental premise that there is only the existence of a thing. There is no 'non-existence' or negative state. There is light with a source, yet there is no source of dark. There is good but no source of evil.

The thing with these pictures is they reside more fully in the sub conscious mind. That mostly unknown powerhouse that runs our lives. This is thought to be associated primarily with the medulla oblongata up to the thalamus, the most basic parts of the brain.

Early man, with his early brain, knew the power of visualising his desired outcome and cave paintings are a testament to the success of the species in that regard. It's still as true today. Ask any of the 'human potential' millionaires and they will tell you to "*see* what it is you want." That is because the sub conscious doesn't see negatives. There is no such thing as 'no apple.' Try and think of that and all you see is the apple.

If we make the mistake of seeing what we *don't* want we get that instead. This is why I refuse to watch those adrenalin movies that depict bike crashes.

So where do symbolic representations come into the equation? It has long been recognised that symbols play a very basic yet powerful part of our life. We use them to represent language on a daily basis, some in the form of letters. Most advertising is using some sort of symbol, be it a brand name or something that sparks recognition in the mind and an emptying of the wallet.

Apart from language, the circle represents infinity and eternity. The ring we use in wedding ceremonies is just that. It says, "forever" to the one we love. The number 8 also has that characteristic only on a different level. The Möbius Strip is one example of an infinite path with only one side. The cube represents three dimensions. If opened out it forms a cross which represents Man. We could fill a book on the subject. In fact many people have, Carl Jung for one.

The Möbius Strip

Taking one arm of the Fire Wheel we see the ancient Nordic Rune, Sowulo. As you can see, various traditions use slightly different representations.

Germanic : SOWULO or Sowelu. The Anglo-Saxon name was SIGIL; Old Norse: SOL; Traditional meaning is the Sun.

The ancient Northern people, like all indigenous peoples, regarded the Sun as a life-giving force.

Sowulo represents the higher will or intent, as well as the sense of self and self-worth. It is the highest force in the Self, directing the individual's evolution along a specific path. Sowulo is associated with spiritual guidance and leadership. It can direct a course of action or state a positive purpose. Success, goals achieved, honour. The life-force, health. A time when power will be available to you for positive changes in your life, victory, health, and success. Contact between the higher self and the unconscious. Wholeness, power, elemental force, sword of flame, cleansing fire.

If, during a casting, you pull this Rune from your goat skin bag, good on ya mate.

If these symbols really do have an effect for change on our psyche, then Sowulo is an incredibly powerful symbol to have about your person. Seeing as the fire wheel is nearly as old as man then it's effect upon us must be interwoven with our deeper self just by association. As with any endeavour, and especially racing, every little helps.

It's still a very popular symbol. I found this Fire Wheel on a toilet roll. Anyone for curry?

Chapter 2
Sponsors and supporters.

Initially, I want to mention Mitsubishi as they were the first, outside Paul's sphere of influence, to take up the mantle. Mitsubishi, as you probably know, are the top turbo specialists in the field. They offered to supply two types of turbo in one, a Variable geometry turbo and any help that was needed.

They also asked Paul for specific readings, Inlet pressure and temperature, outlet Temperature and pressure, Pre turbo, post turbo. In fact, they wanted ten to twelve readings which was not only very exacting but also very interesting. (See chapter Seven.) Paul and the team love graphs and numbers so this would be a doddle.

 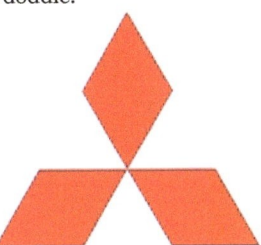

Yes, they built this little number. The number being Zero.

Airbus, the biggest producers of commercial planes in the world, were on board from the start. Having access to their knowledge base, their wind tunnels and some of their equipment had boosted everyone's confidence in this endeavour. Let's face it without them you could have a thousand horses foaming at the bit and still not get anywhere near the record. After two hundred miles an hour, it is aerodynamics and nothing else that makes

the difference. (That is why quadrupling the cubic capacity only doubled the top speed for Glenn Curtiss.) As you would expect, Airbus have that sewn up. This aspect has given Svarog the ever evolving shape that makes it what it is today.

This was turning out to be the very top technology on all fronts. Forget cutting edge, this is the *bleeding* edge in every regard. What a way to start the twenty-first century. It made me wonder what and who would follow. How would we look back on this in our twilight years? With fondness, definitely, and obviously with pride. Also, if you are anything like me, a smidgeon of confusion.

I have always had a hankering for travel. This was going to be one hell of a trip. Unlike any conventional journey, this was going to test the *inner* limits of everyone involved.

Another big name in the aerospace engineering world are Trac. Bob Garlic of Trac, loved the whole project and did whatever he could to help things run smoothly

To those in the 'inner circle,' the hero's of this story are those who believed in what they saw and pushed their hands deep into their pockets. Paul has said that he has been a biker all his life and he wanted the real bikers out there to get involved and be a part of the history taking place. Okay, it would help him but it would also be a buzz for anyone who got on the 'bus' and joined the party. He knows that if you want to enjoy something then you share it and that is what he wanted to do.

Voith was another name from the Farnborough trip. When they ploughed actual cash into the fray then things could really start moving. Paul needed to get a trailer to move the bike around. It was wasting resources to keep hiring a van and then get four or five blokes to lift the beast into the back. Add to that Paul's dodgy spinal chord and it would be a leap forward to get a trailer made.

A lot of the sponsorship money would be needed for the sensors that Mitsubishi had requested the readings be taken from. £4,000 or as near as damn it is one hell of a wedge. *At this point, Paul wants to apologise profusely to all involved for not making the final run.*

Chapter 3
The Team

The only source of knowledge is experience.
Albert Einstein

Success is simple. Do what's right,
the right way, at the right time.
Arnold H. Glasow

This endeavour is not something that you can just say, "Let's have a go at it!" You have to know where your heading and get there a little at a time.

If you have looked into self-development in any shape or form in the last 25 years you may have come across the name Tony Robbins. He has a saying, "Live Life With Passion." It's a great maxim, and sure to bring results to any pursuit that it is leveraged upon. When we put passion into anything we get better outcomes. Even if those results are only the feeling of passion.

Paul Anderson, by his own admission, is not a social animal so meeting him may not be in your future. But let's say for argument's sake that you did, you would see passion at first glance. Getting him onto the subject of motorbikes is not an arduous task, he's already there. Move that subject onto fast motorbikes and he is so far ahead of the majority of us that catching up is a gargantuan task. Why? Because he has passion, container loads of it. And it's infectious, just ask his team. Most are bikers and one, Garry, has been drag racing with Paul for twenty years or so. Another, the aeronautical genius Andreas, had no interest in bikes before Paul pulled him into the fray. All he wanted to do was make planes as aerodynamic as possible. Now he's thinking of getting a bike. I am another that has been swept along on the tide that is Paul Anderson. When Blackjack first mentioned to me that he knew of someone who was going to attempt the record I thought "Oh yeah, okay, sounds slightly interesting." I had intended taking a rest from journalism for a while but just spending a little time with Paul gets your enthusiasm astir. I was definitely along for the ride.

We've all had a bike at some time or other that we feel love for. If we didn't then we wouldn't be bikers. People who get into and then straight out of the 'scene' have not had that experience. For me, it was a Triumph Bonneville. Funnily enough, it was Triumphs and Harley's for Paul also. He started out so far on the wrong side of the tracks to put anyone I know in the shade. Paul had been living in South Africa since the age of seven. At one time he went Awol from the army and found himself starving on a beach. All there was to eat was roasted seagull. So he did. There is a reason why they don't appear on many menus. Imagine leather that tastes of anchovy and you're nearly there. Till the Military Police found him he was on a one-way track that was leading to a meeting with the Grim Reaper. He was a kid that needed direction in life.

Back then, in the 70's, Apartheid was in full and vigorous swing. At the tender age of fourteen, Paul was a white kid, he's still white, and was mixing with a black guy. For some reason, both sides of the divide frowned upon that. But this was no *ordinary* black guy. His name was Joe Mallhas and he was in love with the Harley Davidson motorcycle. He had a huge collection of them. This is the thing with bikers, all social barriers seem to drop. So Paul could be found at every opportunity frequenting Joe's garage. (We know a song about that, don't we?) He would do a few chores and help out and now and again get the chance to learn something about all the Harleys that were stashed there. When things started to hot up with the opposing forces in Johannesburg, Paul decided to move back to Britain with his kids. He had to leave behind many of his beloved machines and start again from scratch. After all, if you have a car full of Secret Police sat outside your house at all hours then it's time to think of No.1 and family. Safety first.

He still loves old Triumphs, but now his attentions veer towards Harley Davidson, in particular the V-Rod. Paul sees so much potential in the motor. In fact he observes it in the whole machine. He showed me a section of the frame that he had cut out to be able to route the exhaust as it relates to the initial turbo set up. The tube has a wall thickness of six millimetres. That is a quarter of an inch to the old timers. If over-engineering makes for great safety then Paul intends to force that status to its ultimate limit. As he says, "The maths works out". So don't for one minute think he's just going to push and hope for the best. Paul has some of the top brains in the industry on this project. Working at Airbus, as he does, he is privy to the apex of technology and the minds behind it. There are no real unknown quantities involved. Even so, there is always nature to contend with. From the nature of steel to the nature of the weather. From the nature of the track to the nature of the fuel. Somewhere along the line, there will inevitably be an unknown quantity. Knowing what I do of Paul, he will take it in his stride and overcome anything that may arise. He has done so many times before. Like a hunter who aims beyond the trees at his prey, Paul has his sights set on the prize. Anything in the way of that doesn't stand a chance.

It can sometimes seem like the equivalent of being a drunk or a junkie. All they want to do is get the next fix. They end up socially alienated and could lose their life. Getting a bike to the sort of place it needs to be to do something spectacular can have exactly the same effects.

The people involved in all of this seemed to me to be making up the Fire Wheel. Central to everything was Svarog, the spindle within the hub. Around him are Paul, Garry, Rich, Andrija, Connor and Ethon. Outside that central Sun are the arms made up of the people not in the team but who still give their time as if they were.  (To Paul, the most highly prized possession a human has is time.) Running off that fire wheel are the purple flames. Purple is a sacred colour in a lot of religions, and the people representing the flames are highly regarded in this instance as they are the sponsors and supporters.

Garry, Paul, Ethon, Richard and Bob.

The Team.

Garry.
From a young miss-guided lad with no direction and no idea what he was doing with his life, he turned into what he is today after meeting a guy called Tommy Morris. He's been into bikes all his mobile life, sold them, rolled them and fixed them. He has gradings in five different martial arts and has a philosophy of life taken from Eastern tradition. With twenty five years married to Fi, he seems content with the way his life is rolling along.

Richard.
Richard is the numbers guy. If that doesn't cover it for most of us then there is very little that I can add that will make a difference. Mathematics was never my strong point, so when I hear stories that these guys actually pass jokes around written in equations then you get an idea just how far above the majority that Rich is.
The best story I have about Rich is the time he broke his leg. He has a couple of old Triumphs and rides *every* day. One day he put the high compression bike away to give the low compression a bit of a run. It had been a while since he had kicked it over and used the same force he would for the other bike. It turned way too easily and he snapped his femur. Great story, but it turns my stomach.

Andrija.
Andrija Ekmedzic, an aerodynamicist PhD joined up in 2008 & applied his skills to the intercooler inlet & outlet ducting. From there his attention went to the 'mould breaking' bodywork including the cool looking tail fin.

I had a chance to talk to a falconer and was told that the Peregrine Falcon has been recorded at 240 miles per hour and possibly up to 270 in free fall. Any object allowed to reach terminal velocity through the force of gravity will reach 122 mph. The falcon can double that. A total miracle of nature and the fastest animal on the planet, it's to be respected by us humans let alone it's prey. So my first question to Andrija, "Did you take inspiration from the Peregrine?" He gave me a quizzical look and said "No." For Svarog he took inspiration and design queues from none other than the Super-marine Spitfire. He got RJ Mitchell's designs that have had the most thorough testing imaginable and worked from there. Oh well, it was just a thought.

Ethon.

At the tender age of seventeen, (When I first met him,) Ethon is the youngest team member. To have this opportunity to get involved at such a young age, even if it's only to keep the garage tidy, polish the bike, take all sorts of shit from the older ones, must be just about the best thing that can happen to someone who relishes fuel powered two wheeled locomotion. But Ethon does more than that and has been found with spanners in his grubby little paws, putting stuff together, using the lathe and more. That takes a lot of trust on Paul's part. Trust that is well earned.

Sebastian Parent

Take one handful of Spaghetti and sprinkle a little devil dust on it and there's your wiring loom. However, Sebastian doesn't cower in the face of Occult Satanic Italian nasty food. He wires up engine management systems for planes and helicopters fercrisake. Walk in the park this.

Connor

This young genius is the son of Paul's partner, Christine. He has a way with numbers and computers and basically the thought process that should be natural to someone well in advance of his years. Another young one with a great future.

Me.

I was very honoured to have been invited to actually join the team and get a cloth patch. I was just expecting to be on the outside looking in, doing the thing I had been doing for the previous ten years or so, and being a photo-journalist. What can I say about me that isn't blowing my own trumpet? Well, I'm not musical, but I do like to use words to get an idea across. Since around 1985 I have known there are stories in me, one is the story about my love for all things mechanical and creative. You get the idea.

Left to right: Andrija, Ethon, Yours truly, Connor, Paul, Rich and Garry.

Photo courtesy of Mick Kirton.

"What lies behind us and what lies before us pale
into insignificance compared to what lies within us."
Ralph Waldo Emerson

Chapter Four
The Bike

"There is more to life than increasing it's speed."
Mahatma Gandhi

2001, and the time of the launch of a new baby from HD. True Blood Harley riders are about to get the shock of their lives. They won't like it, "It's not a proper Harley!" is going to ring out in the 'Old School' quarters around the globe. For a start, you don't need to do anything to this bike to make it do what it should do. The potential to reach 140 miles per hour with enough horsepower to spin up the rear wheel like a muscle bike should.

Not only will the motor set your eyes whirring in their sockets, but the frame it is all wrapped in is modern and functional. This is nothing short of a total revolution in bike building from the Mo Co. But then the design comes from that country that has turned out not only the first ever production car, but is very fond of sausage. Yup, Germany. Stuttgart ("*Das neue Herz Europas*") based Porsche set to with the felt tip pens and A3 sheets of paper to design this bad Mo Fo. Water cooled, safe up to 9,000 RPM. What were Harley thinking of?

Hate Harleys? Think again. This little wolf could turn the tables for those that have never even considered Milwaukee Iron before. How about 115 HP at the crankshaft (100 from the rear hoop) slurping out of the 69 Cubic Inch motor. (1130cc) 100 X 72 millimetre (3.94 x 2.83 inch) bore and stroke. 11.3 to 1 compression held tightly in the cast iron cylinder liners.

Dual downdraught intakes feeding into 53 mm throttle bodies swish the air/fuel mix, readying it for the sparks via a Sequential Port EFI system. Basically, it's going to be the right mix for the circumstances. A number of sensors take into account manifold pressure, crank position, incoming air temperature, throttle position and engine coolant temperature. Then.....BANG. Away she goes.

Both pistons push down on a con-rod spinning the one-piece forged crankshaft. Automotive-style full pressure bearing journals keep things steady while a single crank driven forged steel counter-weight balances out the vibrations. Push all that power through a five-speed gearbox that features a hydraulically operated nine plate wet clutch, and that fat rear tire, through the belt drive, gets its oats.

Specifications:
LENGTH : 93.6 in (2375.6 mm)
SEAT HEIGHT : 26.0 in (659.9 mm)
GROUND CLEARANCE : 5.6 in (142.1 mm)
LEAN ANGLES (LEFT/RIGHT) : 32.0 / 32.0
RAKE/TRAIL : 34.0° / 3.9 in (9.0 mm)
WHEELBASE : 67.5 in (1713.2 mm)
DRY WEIGHT : 595.7 lbs.
BORE X STROKE : 3.94x2.84 in. (100.00x72.00 mm)
DISPLACEMENT : 69.0 c.i. (1130 cc)
COMPRESSION RATIO : 11.3:1
FUEL SYSTEM:: Electronic Sequential Port Fuel Injection
OIL CAPACITY : 4.5 qts (4.3 litres)
FUEL CAPACITY : 3.7 gal (14.0 litres)
EXHAUST SYSTEM : 2-into-1-into-2
PRIMARY DRIVE : High contact ratio spur gear
FRONT BRAKES : Dual 11.5 in (292.1 mm)
REAR BRAKE : Single 11.5 in (292.1 mm)
FRONT WHEEL : 19-inch disc
REAR WHEEL : 18-inch disc
FRONT TIRE : 120/70ZR-19
REAR TIRE : 180/55ZR-18

So that is your standard 'shandy'.

Svarog has very little of those same internal's left. In fact, when Paul was offered a 'Rod' at the Bristol Bike Show he turned it down as there is nothing of any use for Svarog in a standard bike.

Once the truth serum reached my brain I had to be honest, when I first saw the new V-Rod I was not at all impressed. Here was a 'suit' showing off the bike at the Bulldog Bash for the first time. We asked to hear the engine and being more of the vintage, or at least 25 years old, Harley regime I thought he hadn't actually started it. It was emasculated, I want something I can hear. Who would go for this? Harley riders? Nah! Sports bike riders? Nah! Metric bike lovers? Nah! Then who? Well, it seems that prejudices were not on the list of the first folks to snap up this forward thinking machine. And I'm betting they were glad of that.

"There is nothing so stable as change." Bob Dylan.

When someone buys a new Harley they usually know it's a base to start from. Some people are happy to leave it at that. Most of the folks I know buy one with the intention of doing something to it. That may only be changing to decent brakes and carburettor. The V-Rod was meant to be different. It was intended to be a bike that already *had* the performance it was supposed to have. In fact, it did. There is no need for anyone to do anything to this bike to make it what it should be. But Paul is not just anyone.

There wasn't a quantum leap, however, from a standard to the machine in front of us. Before the Black bike that became Svarog, there was the silver bike. A lot of work was run through that setup. Work that could have been seen as failure due to the run of bad luck. Truth is, there are no failures only lessons on how to succeed. If we learn, rest and move on then the path to success is still being walked.

So there was Paul back in 2003 with the silver bike. Known in his local circles as the Yamaharley. But then people were very biased back then. He had spent a not inconsiderable amount 'tarting' it up. Drive pulleys, wheels, all sorts of shiny do dahs. The silver bike was one of the first around. Paul customised the hell out of it, and no one noticed. "Oh," they would say, "So that's what a V-Rod looks like!" Um, no. That's what *Paul's* V-Rod looks like.

'Dragging' it had become a must if only to show that it was different. He went the serious route and ordered a couple of grand's worth of turbo. Paul put an *afternoon* aside to fit it and a few days later had a working set-up. Fifty horses more, all ready for the road. A melted piston ensued but it was still chucking out one sixty BHP with that rattling around inside it.

The piston has ended up on the wall as an offering to the "God of Speed." A traditional artefact in the drag racers stockpile of memorabilia. After a total rebuild, it was found that a seal had a bit of dirt and two drops of oil came out. Another re-build and another race and it was melted again.

That is when "Black Bart" Turned up. The black bike was Dragging in the States with under ten second quarters or thereabouts and was a fully fledged workohorse.

But Paul bought it with all the trick bits on it but a blown motor. He had it rebuilt to his spec for a turbo set up. The fatal mistake that Paul and Garry had made was letting the Americans touch it. Supposedly one of the best drag racing engine builder in the States at the time.

If you take into account this motor was put together by a man thought to be at the very pinnacle of the drag scene, then the story that unfolds may surprise you. On arrival on British soil the engine, for all intents and purposes, looked healthy and ready to roll. Paul and Garry had been setting up the rest of the bike and the turbo was ripe to squeeze out what it could. A lot of hard work and late nights had gone into getting it qualified to run at the yearly meet at the Bulldog Bash in 2005. Every night till midnight for months beforehand. Even so, when all is said and done, sixty-eight miles an hour and absolutely no horses is just a let down.

Inspection brought up a silver looking oil. This is not good news in any circumstances. Paul was devastated and sank into a pit over the fiasco. So much so he didn't even tell the supplier of the engine what a wanker he was. (I'm only telling him now so he has a chance to get his sorry act together.) The next ten months saw nothing happen to the bike at all. Paul just couldn't face it. Eventually the drive in Paul fired him up again and he was soon back into the engine. The silver turned out to be an exuberance of a sealing compound that the Yanks had used. The front head wasn't tightened down at all, the valve timing was twenty eight degrees out. Bolts had been snapped, a damaged cam chain tensioner bolt had let the tensioner come away from the body of the motor which let the oil pressure drop. When the gearbox main shaft had been put in the assembly had been tightened without positioning the pins properly, this made the whole gearbox out of alignment.

Paul stripped it all down and did it himself, it was the only way to be sure.

After a while drag racing became, well, a drag. Too many regulations and people telling you what to do. The fun was going out of it and the bike was put to the back of the garage. Knowing that was sacrilege, what are you going to do with a bike that could hit two hundred miles an hour? It didn't take much alcohol to churn up the idea of going for the land speed record. From small acorns and all that.

Chapter Five
Theory and Science

"Always bear in mind that your own resolution to succeed is more important than any other."
-Abraham Lincoln

Or as the Stoic philosopher Seneca said:

"It is not because things are difficult that we do not dare. It is because we do not dare that things are difficult."

As the project progresses, the amount of available technology grows along with it. Advances in technology usually run on an even keel with the evolution and growth of the human brain/mind. With the development of any structure, a human relationship, a multi billion dollar corporation or a tomato plant, there is change. It's obvious really. As change occurs a wave is formed, the crest is order and the trough is chaos or entropy. The fluctuation between the two has been working in the universe for 13.73 billion years. Give or take a few days. If a structure, organic or inorganic, grows too quickly the chaos principle reigns. Out of chaos comes order. (Orderliness is allowed to come about under certain conditions such as slower growth. It is the nature of the universe to seek it on *every* level.) At a certain stage of disarray there is a bifurcation point, a split. The structure can either evolve in one of an infinite amount of ways or it dies. This project has been allowed to grow and develop at a steady pace. Paul knew from past experience, if it grew too quickly then motors would suffer the effect of total chaos and blow. Been there, done that.

His early days of drag racing were a sharp and sometimes expensive learning curve. With Airbus stepping up to the plate, all that changed. If you can map out where the engine is on a computer then the steady and almost predictable growth can be foreseen. Everything has to be taken into consideration. The thickness of the cases around the big end bearings, the temperature of the explosion, the ability of the pistons to cope with that temperature. It can all be put on the table, examined and even run through in a virtual sense to see how things are going to pan out. This has to be a good thing when the engine alone has in excess of sixteen thousand British pounds supporting it.

When Mitsubishi came on board they asked for readings in ten to twelve specific locations so that a custom blower could be made for Svarog.

Once Paul had got all the shows of 2010 out of the way he needed to strip the bike down and find ways of fixing all these sensors in place. With the wiring loom, um, looming, October was going to be a full-on month of high focus work. If only he could fit another sixty days into that month things would be just dandy. When anyone squeezes more days into a month it turns out that that time scale eventually has to go by another name. The name here is December. As you may know, if assembling mechanical things falls into your sphere of pastimes, nothing ever happens when you want it to. Building *anything* takes twice as long and costs twice as much as you first think. Even if you take that fact into consideration the same rule applies. There is no escape from this universal law.

I don't know about you but before I met Paul I had no idea what an Intercooler was. I know Eddie Stobbart is very fond of them. Man, Daf, all the big trucks would advertise that fact. But wtf does it do? Cool stuff, I hear you say. You're right, turbo workings to be exact.

Paul looked at the car and truck manufacturer and saw they mounted these things in front of the radiator. He did the same with Svarog which turned out to be absolutely useless. The intercooler picks up heat from the radiator, the air can't go through at a decent rate. All sorts of problems arose so it was moved and a duct was made to funnel the air in. That would be okay on a road going bike but not one that's designed to trundle along at over two-hundred mph. Now, this doesn't just relate to Paul's set up but to *all* makes that have some sort of temperature reduction system. Oil, water, you name it and it will need cooling at some point.

What was found in the early wind tunnel experiments was that the fastest that the air could go over the elements of a radiator was around sixty miles per hour. Go any faster and the air just backs up. So that is where Andrija came in with the design of that reverse shaped funnel. See how it gets bigger towards the back and wraps around the wheel?

Imagine yourself as a ten-year-old. You know, when you first farted into a bottle to set light to it. That was where the idea came from. Doesn't that make you feel safe when you see thousands of tonnes of Boeing go over? All tested out by farting in a bottle.

Seriously though, back to everyone else's reality. He actually took Mitchell's Supermarine Spitfire cooling system and looked into that to invent this system. So, it has its roots in one of the most iconic planes of the 20thCentury.

The wind tunnel tests, the experiments for Mitsubishi to get their *thang* together and the wiring and imminent firing of Svarog all started to come together around Christmas 2010. The tunnel tests would put the tail fin firmly on the menu. The data logging equipment was lined up to arrive in the middle of October. Temperature sensors were relatively cheap at £70 a pop. Pressure sensors go for a little bit more, £400. Blimey! The minimum Paul needed of those was three. Like Garry said; "You know the old saying, there's no such thing as a free turbo."

Everything *was* lining up for December of 2010. The wind tunnel was 'go' so Paul knew he needed to get the bike ready to roll. The Mitsubishi experiment, the rolling road to dyno-tune it and the wiring all needed to be brought into alignment. The new wiring loom was the catalyst at this point. Then the whole bike had to be ripped down and re-built to get rid of all the show stuff and put the serious stuff on. New porcelain wheel bearings, proper fasteners, the works. The spring of 2011 was looking to be the schedule of the first run. For the first time in the history of the bike Paul had some funding behind him so events were in line to happen. But when you look at £500 for a set of pistons and £2,000 for a crank then being a little frugal with the funds makes good sense.

At least, that is how it was panning out back then. But like we saw earlier, things never go as you plan. It wasn't until September 2011 that this point was reached. No matter how close you *think* you are you may find that the view is through a telescope.

Wind Tunnel.

Centre to Centre.

What's all this Wind Tunnel chat about?

To get a streamlined vehicle to operate within acceptable tolerances and stay on a straight course, two centres have to be observed.

We all understand the 'Centre of Gravity' although I thought I may be a *little* unsure so checked it out and found this on a website.

$$cg * W = g * SSS \; x * rho(x,y,z) \; dx \; dy \; dz$$

Now I am *more* than a little unsure.

But what about the 'Centre of Aerodynamic Pressure'. This is a *theoretical* point where all aerodynamic 'moments' occur.

There is an interaction between these two points, especially in motorcycles and aeroplanes. The centre of gravity *must* be ahead of pressure. By a big margin if you want to move in a straight line. This brings to mind the Tea Cups at Alton Towers, what frightening stuff. They are forever overtaking themselves, you can count me out of that ride, baby!

Beyond a certain distance apart then it will always be self-righting. The tail fin will help to keep this scenario together 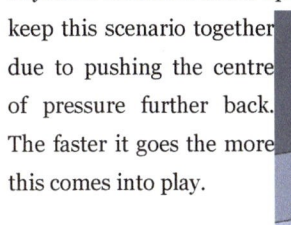 due to pushing the centre of pressure further back. The faster it goes the more this comes into play.

So the bike, including 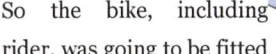 rider, was going to be fitted to a small plate about the size of a couple of boxes of cornflakes. That would, in turn, be mounted to the calibration equipment that is about the size of a double-decker bus.

Usually, only models of things go in this tunnel as getting an F1 fighter jet in can be a little bit of a struggle. Even with one bloke on each corner.

With wind speeds as little as three hundred miles an hour (Little compared to jet fighters,) the whole unit must be very well secured. Anything over one hundred miles an hour is enough to lift a fully grown man off his feet. Paul had to be bolted down if he didn't want to be flung to the back of the chamber. And yet people pay money for that sort of thing.

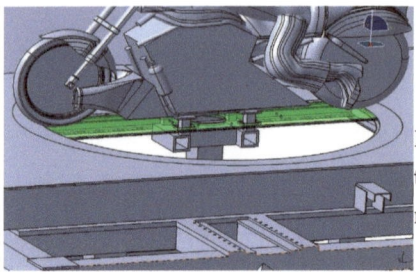

Testing.

I finally get to see what the inside of a wind tunnel looks like. Much like you would expect really. It's a tunnel with wind going through it.

Ah yes, but what a groovy wind it is.

I've been privileged to have been in a lot of select environments over the years. Clubhouses, backstage, track-side and even in the presence of (almost) Royalty etc.* etc. So when I got an invite to a precision process like the Airbus wind tunnel I was very grateful for the opportunity.

(Some I haven't lasted very long in as I was improperly dressed.)

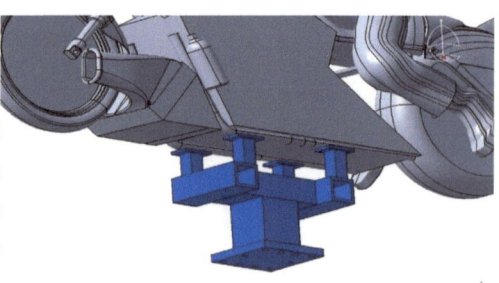

I know my idea of a wind tunnel involved smoke streams. After all, it's the media's conceptualisation of the process. But that is pure BS. There is no data to be gained from that. Unless you are trying to see how long a joint will last at a Rasta party. In a real wind tunnel, all the measuring and information collection happens under the floor. The gantry that suspends the object to be measured is, as already mentioned, the size of a double-decker bus. I didn't get to see that as it's hidden.

Usually, only three or four people are in the control room at any one time. I would guess that over the course of the day we had upward of fifty visitors. Either popping in for a quick peek or staying as long as they could. Including the head honcho of Airbus who came down from Edinburgh. It's not often that we see someone doing two hundred miles an hour on a bike. Even rarer is doing it while not moving.

Whenever we think of the greatest of human achievements such indicants as the Great Wall of China, the space program, the industrial revolution or even the invention of gunpowder, logically flood our awareness. Although most "Great Occurrences" in modern day thought are manufactured by a throng, every now and again it is down to a handful of entrained minds and random scribblings on the back of a fag packet. This day saw computers instead of paper doing the heavy work.

Just how useful is this information going to be? In the real world, air is never in a constant state. Chaos theory plays a part and is a testament to that fact. However, the data will be close enough to tell us if the prototypes are in fact doing their jobs.

Talking of *'coordinated'* pandemonium, that's how it seems to be up until now with chaos being taken in the scientific form. That is all about to change. When Paul glibly mentioned to Garry in 2004 that he wondered what would happen if he put a turbo on a V-Rod, he could not have envisaged what was going to happen over then next eight to ten years.

If you have between £75,000 and £125,000 burning a hole in your dungaree pocket and want to know how you would fair on the surface of Saturn on a 'light breeze' day then the only realistic solution is to book yourself into a wind tunnel.

I can only guess at Paul's perception of the day. There is nothing to give you the sense of moving, yet hanging on for grim life while doing the equivalent of Mach 0.16. He said that looking at the black screen with the throat of the tunnel opening out before him and the wind pushing him back was a very "down the rabbit hole" experience.
The first test was on a naked bike. Then he was strapped on with a 'fall arrest' harness yet was really quite nervous about the whole procedure. Jokes about camo brown trousers were rife. Mark, the safety officer, didn't have the same reservations. From his point of view, all seemed to be fine. After all, they tested the apparatus with an expendable person. Jeremy Clarkson had already been in there. But at a much lower 130 mph and with less amount of interested parties attending, obviously. Actually, I don't know the truth of how many watched him, but if you know anything about Clarkson you'll know why I stand by that statement as none of us really care.

From the outset, things get a little beyond my ability to describe. So I will leave it to the experts as in Richard's report and the photo's. It should give you some idea on a level you may appreciate. Over to you, Rich with the official report:

Introduction

On 21st December 2011 an extensively modified motorcycle was tested in the Airbus low speed wind tunnel at Filton. The machine is privately owned and financed for the purpose of land-speed competition. The testing completed one of many important phases in its development program.

Test Objectives

The test objectives were essentially two fold. Firstly the aerodynamic performance was progressively investigated, starting from bare motorcycle and finishing with a final configuration of rider with additional aerodynamic devices. Secondly the aerodynamic performance of a bifurcated intercooler for the engine's turbocharger was explored. Observations regarding the ergonomics and work load of the rider were to be made where appropriate.

Test Set-up

The machine was mounted on a four pillar mechanism that was in turn connected to the wind tunnel's six axis balance, allowing measurement of each orthogonal force and corresponding moment. A floor plate had been fabricated that allowed the yaw to be swept from -3 degree to +3 degree. Follow an appropriate safety review the tests with a live rider were conducted with a safety harness that in turn was tethered to the machine. The rider was provided with a "dead man's handle" so that the test run could be terminated if conditions became uncomfortable. The set-up had been assessed and validated as being safe. The forward and aft surfaces of the intercooler were instrumented with an array of pitot tubes.

Test Procedure

Each run with 0 degrees of yaw started at 100mph and was increased in increments of 10 mph up to 200 mph, unless first terminated by rider or if stabilized wind tunnel speed could not be achieved. Measurements were taken once the wind tunnel speed had stabilised. The effects of yaw for the particular configuration were then measured at a constant 150mph, sweeping in ½ degree increments from -3 degree to +3 degree. The test configurations were as follows:

1) Bare naked motorcycle

2) Motorcycle with rider.

3)   Motorcycle with rider and nose fairing

4)   Motorcycle with rider, nose fairing and front wheel fairing

5)   Motorcycle with rider, nose fairing, front wheel fairing and rear aerodynamic device

In this photo are John Adams [MD] and Bob Garlic both of Trac.

(I didn't get a picture at that stage with Paul astride as I was videoing instead. Just in case...)

6)   Motorcycle with nose fairing, front wheel fairing and rear aerodynamic device

Notice the use of gaffer tape. I was also relieved to see a horse shoe in the control room.

I managed to get permission to show the layout of the tunnel. I have to say the whole thing was not what I was expecting. RL.

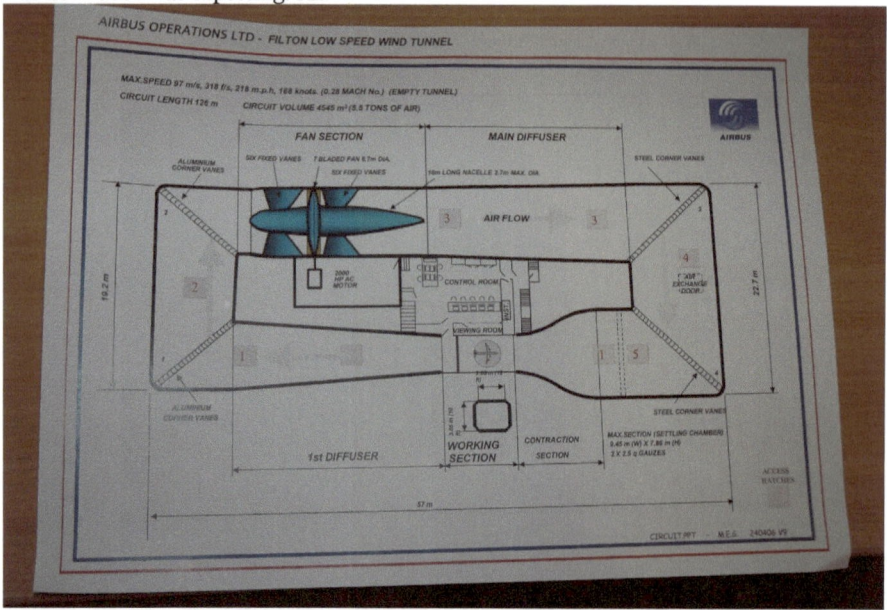

This picture was taken from the contraction section with Garry and Ethon both doing their own particular form of yoga.

Qualitative Observations

With the bare motorcycle and no rider there was no visual evidence of significant vibration until about 150 mph. Vibration started on the intercooler hose. At 180 mph it had progressed to the seat unit and chain guard. In the bare configuration the rider experienced difficulty in holding his head down at 140mph, eventually terminating the test at 170mph on his own initiative. With no fairing to shield the rider's helmet from the airflow the uplift generated by the air vortex had become unbearable. Addition of the nose fairing made things more comfortable for the rider, making 200mph tolerable. However the rider found it progressively more difficult to keep his head down at speeds past 150mph. The resulting effects on overall drag are discussed in the Quantitative Results section. On final observation was that the rider's leathers were inappropriate for the task. At 160mph the jacket began to billow and the trouser legs began to flap.

Quantitative Results
Drag

Straight-line drag forces were compared using a dimensionless *Coefficient of Drag* (*CD*) derived from the configuration of Motorcycle with rider. Recall that tests with the rider on the bare machine were terminated at 170 mph. Drag data beyond 170 mph will be discussed separately.

The data up to 170 mph was analysed for the following configurations:

Bare motorcycle with rider (*Bike Rider*)

Motorcycle with conventional handlebar fairing and rider (*Bike Rider Fairing*)

Motorcycle with conventional handlebar fairing, bespoke front mudguard fairing and rider (*Bike Rider Fairing Fender*)

Motorcycle with conventional handlebar fairing, bespoke front mudguard fairing, aerodynamic device and rider (*Bike Rider Fairing Fender Device*)

(See Attachment 1: Motorcycle Test Configurations)

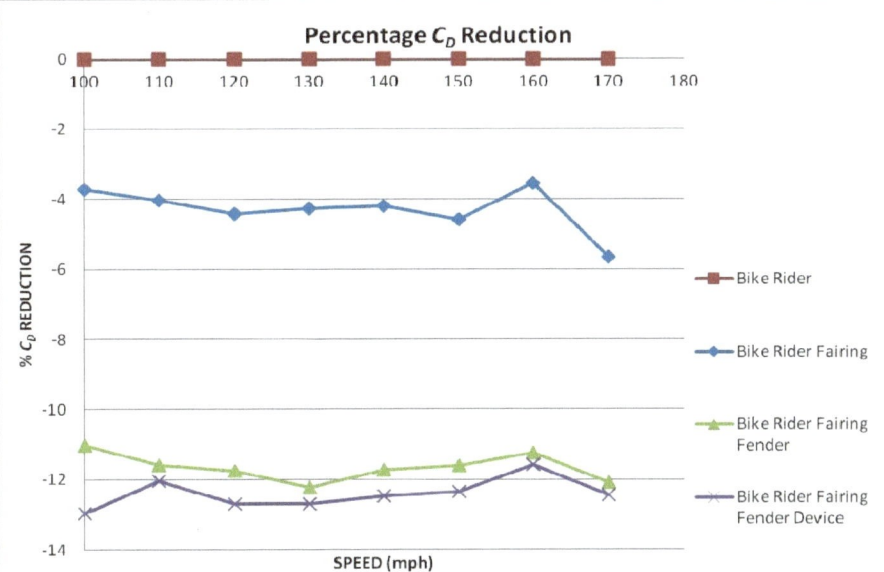

It can be seen that the bespoke front mudguard fairing had the most pronounced effect on drag reduction. In comparison, the conventional handlebar fairing reduced drag by only half as much. The aerodynamic 'fin' device produced a marginal reduction in drag, but its primary design requirement concerned Centre of Pressure (to be discussed later).

Rider Effects on Drag

The percentage increase in drag was calculated from the CD (Coefficient of Drag) derived from the configuration of Motorcycle with rider. The results are plotted below:

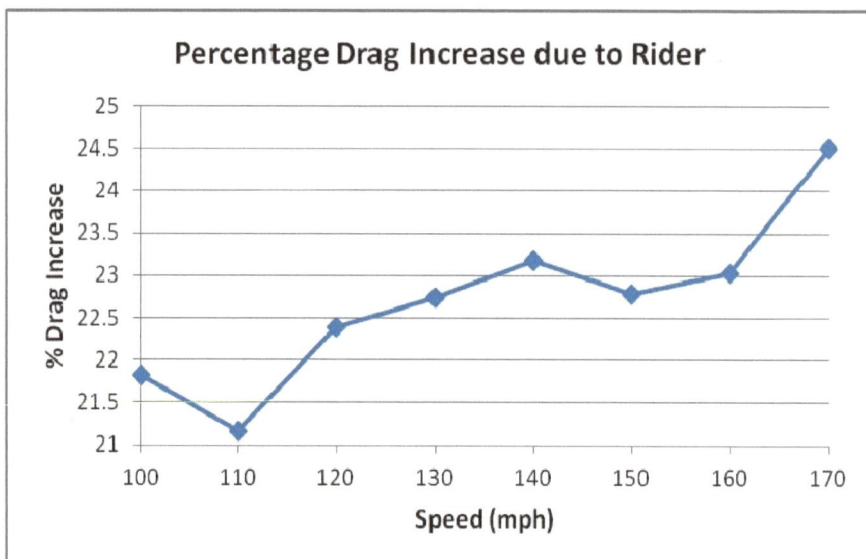

This analysis has limitations. The range is limited because the highest speed achieved with the rider on the bike was 170 mph. CD was derived from the frontal area of rider on the bike. Despite this the results show an interesting trend of rising rider induced drag from 150 mph. This certainly justifies the rider's decision to pull the "dead man's handle" at 170 mph. It must have been becoming progressively more uncomfortable for the poor guy!

There is a general trend of rising drag that could be partly reduced by the rider wearing one-piece leathers that didn't flap and billow in the wind. However there is a residual drag that is apportioned to the front area presented by the rider. Reducing this area would increase the machine's top speed.

Currently the machine is configured for the rider in a horizontally prone and feet back drag racing position:

Theoretically this could be improved by moving the foot rests forward and narrowing the handle bars. This would reduce the protrusion of the lower legs, boots and forearms into the air flow by allowing articulation of the hips, knees and elbows. A potential reduction in the frontal area presented by the rider is shown below (shaded region):

Maximum speed is inversely proportional to the cubic root of the frontal area. The area reduction described above would result in an estimated 2% increase in maximum speed.

Centre of Pressure *(CoP)*

Yaw moment and side force coefficients were used to estimate the *CoP* position on the length of the motorcycle. The configurations "*Motorcycle with rider, nose fairing, front wheel fairing and rear aerodynamic device*"- (*CoP* with device) and "*Motorcycle with rider, nose fairing and front wheel fairing*" – (*CoP* without device) were analysed. The results are shown below:

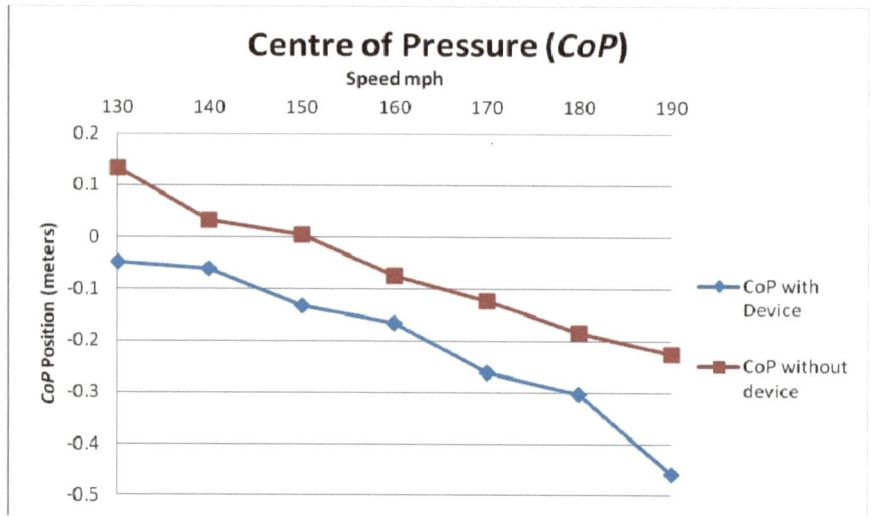

(The 'device' spoken of here is the tail fin. Until recently this was deemed top secret due to the effect it has on not only drag but yaw. You can see a version on the wing tips of modern planes. RL.)

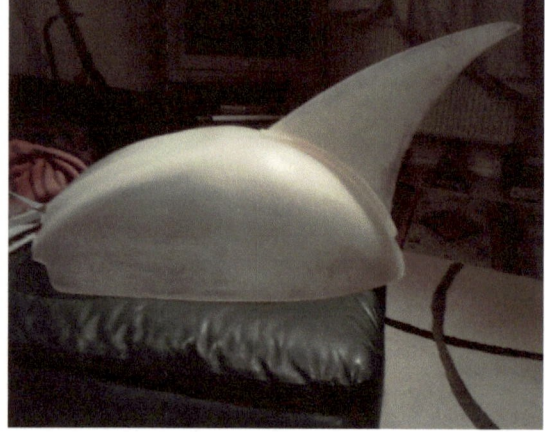

The *CoP* positions obtained from testing with rider and handle bar fairing, front wheel fairing and aerodynamic device have been marked on the motorcycle's length:

CoP Positions on the Motorcycle

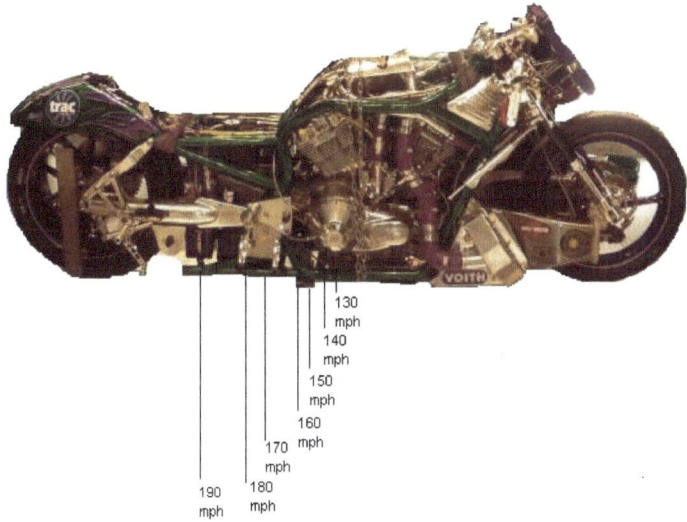

130
mph
140
mph
150
mph
160
mph
170 mph
mph
190   180
mph   mph

The results show that the once above 120 mph that the motorcycle is inherently and progressively stable at increasing speed even without the device. This effect is amplified by fitting the device and has the additional benefit of a slightly reduced drag coefficient. This demonstrates that the engineering judgement behind the initial selection of the device's surface area was correct – large enough to have a pronounced effect on pushing the *CoP* back with increasing speed, but small enough not to have an adverse effect on drag and in fact contributing to drag reduction. The aerodynamic device will ensure progressive straight line stability at increasing speed.

This picture provides an illustration of the device in it's first embodiment as it came out of the 3D printer. This was before being made of carbon fibre and attached to the rear of the bike, as you can plainly see.

It was made in two parts. The fin slides into the guard. When it was made with carbon fibre the same system was utilised.

Effect of Front Wheel Fairing on Side Force

The effect of the front wheel fairing was investigated by analysing the side force coefficients measured at a constant 150mph while sweeping the machine through yaw angle ½ degree increments from -3 degrees to +3 degrees. Two configurations were considered, Rider with Nose Fairing and Rider with Nose and Front Wheel Fairings. Air density was taken as 1.225 kg m-3. The results are shown below:

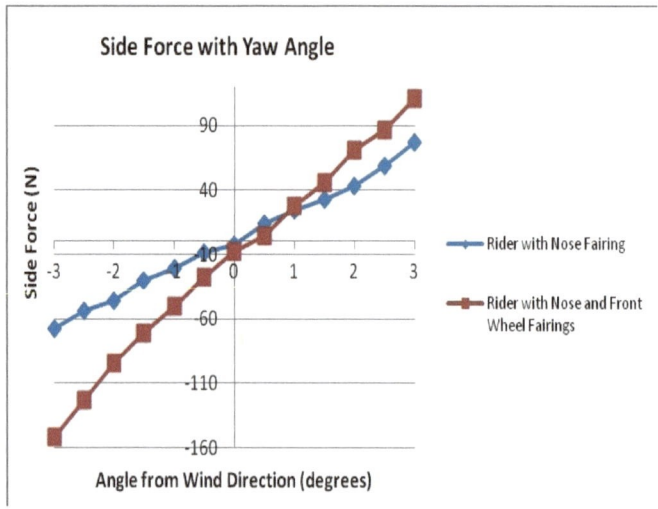

With negative yaw angle the side force is more than doubled with the addition of the front wheel fender. With positive yaw angle the side force is only increased by approximately 50% and only when above + 2 degrees. This effect might possibly be attributed to the asymmetric configuration of the bike, in particular the turbocharger inlet mounted on the left-hand side of the machine. The bike is currently being modified by replacing the existing turbo with another type. The corresponding inlet configuration will offer less asymmetry.

Performance of the Bifurcated Intercooler Duct

The front face of the intercooler was instrumented with 13 pitot tubes and a further 13 were placed on the rear face in corresponding positions. This allowed the pressure drop across the intercooler to be measured at various locations. The results are shown below:

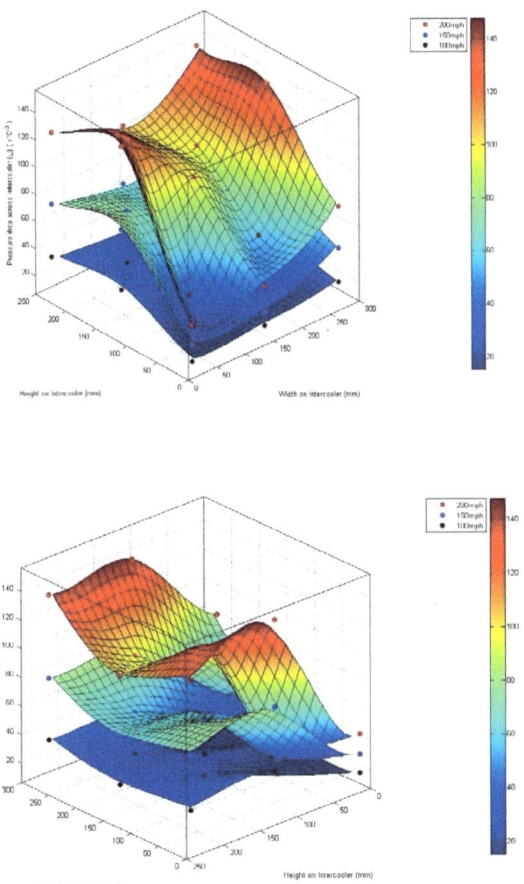

The graphs above show the pressure drop across the intercooler at speeds of 100, 150 and 200 mph.

The modulation due to the bifurcation can be seen clearly. The "valley" in pressures can be attributed to the "shadow" from the bifurcation.

Corresponding velocities at the immediate face of the intercooler were obtained using a nominal air density of 1.225 kg m-3. The results are shown below:

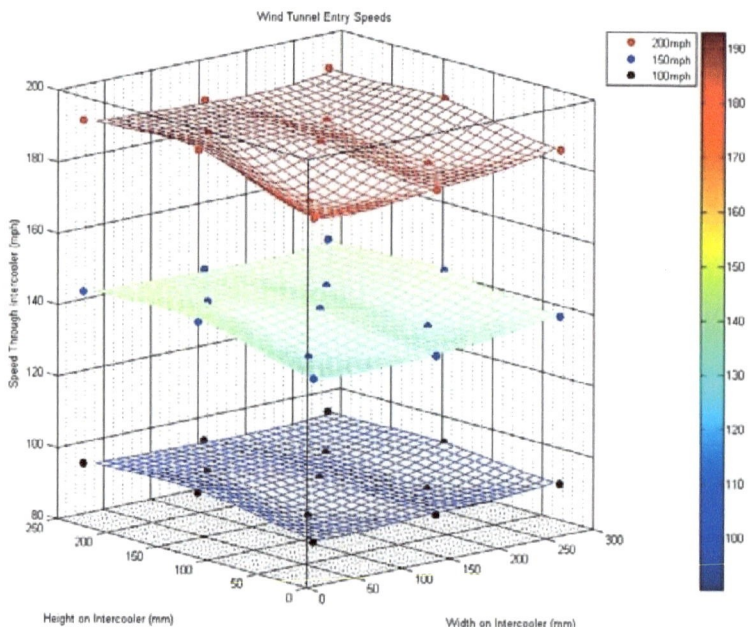

Exit velocities from the immediate rear of the intercooler are shown below:

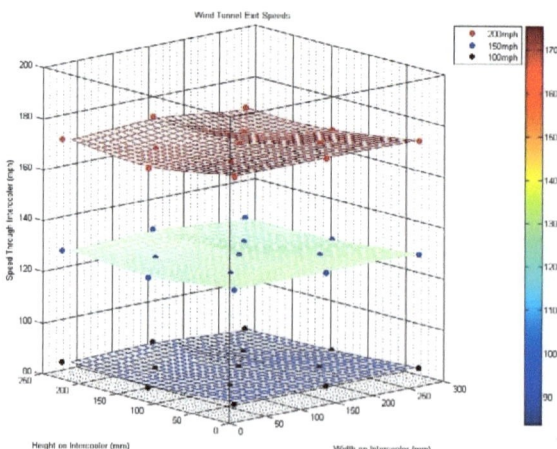

What is more telling is the drop in flow velocity through the intercooler as shown below:

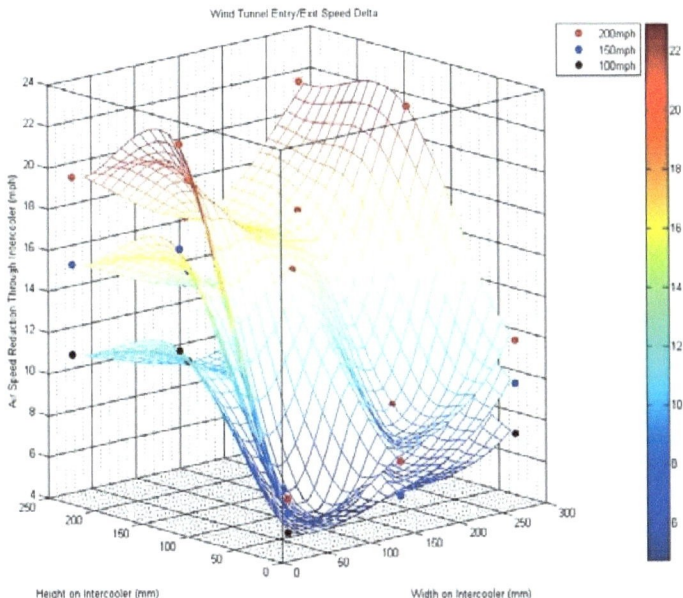

The drop in velocity is minimal at the lower edge of the intercooler. When this is considered in conjunction with the corresponding minimal pressure drop it suggests that there is very little resistance to air flow in this region.

In general the results show that the maximum heat will be extracted from the upper left and right-hand regions of the intercooler. The more uniform distribution of air flow velocities out of the intercooler shows that the heat extracted will be carried cleanly away and exhausted out through the duct's exit nozzle.

The bare machine – note the bifurcated inlet for the turbocharger's intercooler:

Machine ready for rider and in full aerodynamic configuration – note that the front wheel was removed for ease of test rigging

As you can plainly tell, I couldn't have explained it better. Okay, I couldn't have explained it at all even though I did understand it. Most of it. I just hope you can get something from it.

Chapter six
Philosophy

When encountering anyone on any level we find that they have a certain set of rules or guidelines that they live by. Very rarely set in stone, these guidelines fall under the banner of 'Personal Philosophy.' You don't have to be embroiled in Plato or Descartes to have a philosophy, you just have to have a way of dealing with life. "Shit happens" is a favourite amongst the masses at the moment, and will probably be with us till a better acronym comes about. (That or a better government. Fat chance, no such thing.) Paul's view is very similar, "Anticipate the best, but be ready for the worst." "Hope for everything, expect nothing." Not a bad one to live by unless you see 'hope' as the devil in disguise that it really is. (More on that later.) In the Eastern Yoga traditions, it would be likened to Karma Yoga or detachment from outcomes. Being detached means you never get let down, even if the worst happens. From the start, Paul knew that could easily be a scenario that would play out at some time. Moving a solid through the air at great speed is not without its pitfalls. No matter what precautions are taken, there is always the unknown and the unpredictable.

The Buddhist view is, "What is, is!" The Christians would say, "Lord grant me the ability to know how to change the things I can, the strength to accept the things I can't, and the wisdom to know the difference."

I like all these as they lead to an easy life, and who doesn't want that?

Looking at where that philosophy falls on a record breaking attempt it's easy to see that no part of the scenario is outside the sphere of philosophical influence. If Paul could get no sponsorship from any quarter, (Highly unlikely but still in the realm of possibility,) then he would have to take the stance of realising he can't change that and just get on with it all himself. The worst that would happen is that it would spread over a much greater period as he would only have his wages and the kindness of others to work with. But he would've just knuckled down and got on with it. There is no point getting upset as that just makes things harder.

Then there is getting parts. If what you order isn't what you get then you either get pissed off or you get on the phone and straighten it out. Some things you can change and some you can't, simple as that. As the Hitch Hikers Guide to the Galaxy states, "Don't Panic." All too often we get the urge to hurl the part we have just turned up on the lathe as it is two-thou under-size. But testing the aerodynamic qualities of a bearing spacer doesn't get the job done. The feeling of "that'll show you..." as you let it fly, doesn't last long and then finding the piece at the back of the workshop in a worse state than when you started is not only more frustrating but a waste of time. And it's not very 'Zen'.

Where the philosophy really comes into its own is right there on the tarmac. After all, that is where the action is. It can all go one of three ways. Either you break things by blowing up the motor or worse, dropping it. You don't have what is required and don't take the ticket. Or last, and the one that I hope we are all in agreement on, You take the record with such a high margin that sees anyone in the next ten years having a real hard time getting anywhere near the first rung of the ladder.

Getting emotionally involved in the first two will bring only pain. Sure, there are bound to be disappointments if they arise but the trick is to let them go as soon as humanly possible. That way "getting back on the horse" is not drawn out and success is one step closer.

Alongside philosophy, in many quarters, rides meditation. Meditation is basically removing distraction and being focused. It's a mind thing that brings its own rewards. Building Svarog has been a meditation for Paul. He would let no distraction keep him from getting the bike in the best possible state. Plus he has been so totally focused on the goal that he has almost become one with the whole project. Beats sitting in a cave for thirty years looking at the tip of your nose and eating nettles.

We've all experienced the benefits of meditation at some point. It is the feeling of enjoyment. A feeling of being totally caught up in what we are doing. Art can give that to you as can welding and even soldering. When the concentration focused on the task cuts out all distraction, time seems to become non-existent. We get 'into' the task and go with the flow. We are in the 'zone'. It's something that we all aim for even if we don't know it. Yet it can be achieved with a little discipline by sitting quietly and thinking only of our breath. You don't have to be a hippy to benefit.

Anyone who has ever raced knows first hand what it is to meditate. Even if that is not what they would call it. The pure unadulterated concentration required on the track is second to none. It is focused mind with no other thought outside what is needed in the moment. If you want to know what it is to live in the moment, without past or future, racing will give that to you.

True meditation is described as relaxed concentration. So racing isn't *pure* meditation utilising both body and mind. Let's face it, the body is nowhere near being relaxed. However, it is as near as you can get in the mind state without sitting in the Lotus position. In everyday existence, racing is probably as close to Nirvana as mere mortals will ever get.

Chapter Seven
Building Up to It

Within minutes of meeting Paul, I knew there was the passion involved. The bike was in its drab pre-Svarog trim but still looked like it meant business. Paul went over what had happened so far and how it was just about to be stripped for a little... tarting up. The paint was laid on by Tim Cox and that turned the temperature up to 'stunning.' Just in time, Paul and his ever-faithful crew got busy and put Svarog back together ready for...

Farnborough

So Svarog came out into the light for the first time in five months. He was, (And Svarog is definitely a 'he' unlike our run of the mill bikes that get to be a 'she,') newly painted and had that air of speed all ready to be utilised. A week later he's in the back of a truck zipping up to the Farnborough Air Show. Paul spent a week there working his vocal chords to the... well, they don't have a bone but you get the idea. With hundreds of folks interested in the project, Paul was certain of picking up a sponsor or two. Now there are a few folks out there who would like a shot at this title, but I would hazard a guess that not many of them have the passion and know-how that Paul has garnered. After all, everyone is different. This aspect of Paul shines through and anyone with the interest, and is willing to listen will be hooked by his enthusiasm and know he is the real deal. And so it was with those meandering around the place at the Air-show.

The first, before the gates were even open, to show any interest were a likely couple of lads who looked like they had just drifted in from a weeks tour with a rock band. Wearing shorts and unshaven you might have passed them off for just a couple of bums who snuck in the back way. Even the first few words went along the lines of, *"It's only got two wheels then?"* But within two minutes Paul knew he was talking to two of the top guys at none other than McLaren. Not a bad start to the week. Add to that everything from billionaire's and oil sheikhs to the car manufacturer, Cosworth, pluss many other very helpful folks in the industry of moving things fast, and you can see why Paul came home with hundreds of cards. Of those he put sixty aside that he thought would be serious about it.

First to actually send the confirmation e-mail was Mitsubishi. I was driving home from work when Paul phoned me, excited as a cheerleader in a wet T-shirt competition. They had said they "would help in any way they can and send him some turbo's." Bloody hell! I smiled all the way home and I'm not even going to ride the bike. I wouldn't mind borrowing one of the turbo's though. I've got this scooter...

Now think about this; Svarog, although heavily modified, is in the early stages of this project. Mitsubishi aren't the sorts to piss in the wind and hope some of it cleans their boots. They will have checked out the potential for this and made their decision on their findings. True to their word they then asked for the readings so they could get jiggy with it and produce custom made turbos. How cool is that?

Bulldog Bash
12th to 15th August

First to be counted amongst interested parties were the organisers themselves, the Hells Angels. They 'know' bikes and they wanted to sponsor Paul. Over the weekend, Svarog got many admiring glances and a lot of sincere comments. Would the folks go to the website and become part of the story? A select few did.

Bristol Bike Show
21st August

Being the third week in August, this show can go either way. Rain or scorchio. This year, although cold and overcast it all managed to hold off until it was time to put Svarog back in the van. Just got him in and whamo, down it came. But Svarog was going home with Best Paint, (Well done Tim.) Best Engineering and top honours, Best In Show. A pretty mean feat with five hundred other bikes lining up for a shot. I may be wrong, but I don't remember anything in the history of the show taking home three prizes before.

There were meetings to be had and have them we did. Here are a few notes from the noted ones I made notes on.

Thursday was official team night. We managed to squeeze one out a week.

Paul turned up at my place for our first ever official book planning meeting. This book. His sister had just phoned him to wish him a happy 55th birthday, he had forgotten. The funny thing is, it was my wedding anniversary and I *hadn't* forgotten. How does that happen?

We talked mostly about what it takes to go fast. Any way, it's all about the ratio between the centre of gravity and the centre of aerodynamic drag. If you know the drag you can predict the gearing and horsepower for a given speed. There is no guessing involved.

It was interesting to note that John Noble was getting ready with Bloodhound to also go for another record. He was aiming at one thousand miles per hour on the Salt for the outright World Record. Work on that was happening not ten miles from us. We live in interesting times.

Next meeting. This night would blow me away. Touching on the fuel delivery and aerodynamics, Paul jumped right into the Variable Geometry Turbo. This can give a wider rev range than normal. It is based on the old Shorox turbo design and should easily help an engine up to ten thousand revs per minute. From an engine that was only supposed to get to seven before red lining, that would be some achievement. But then that was what was needed for the record.

The scenario of the British record would be;

Accelerate for 750 meters before hitting the beginning of the timed kilometre. If the bike isn't hitting two hundred mph or more, then it would be a struggle after that. The aim is to be at least two hundred and fifteen. Once in the timing zone, Svarog would need to be able to keep accelerating and reach a top speed of two hundred and fifty miles per hour. The average across the Km is what is recorded. Then it's another half mile or so to slow down to a stop. All being well, Paul and trusty crew will turn the bike around, change underwear, breath out and do it all over again. Within an hour.

Simple, huh? Anyone who has drag raced knows that if you want great acceleration then you have to sacrifice top end. If you want a great top speed then it needs to be geared in such a way that your acceleration is pants. Finding that middle way is crucial to winning. But the middle ground has to be wide as the best of both is the only thing that will even get near to a record. It is the 'Holy Grail' of racing.

Going from my own experience, which I have to say is at the most basic level, it's easy to see how this works. When I first built my Shovelhead chop it struggled to get up to motorway speeds without wringing the revs out of it. Acceleration wasn't bad for a 1978 V-twin but I knew it was hurting the motor. All I needed to do was go up three teeth from (23 to 26) on the front sprocket to overcome the problem. Simple. At that point, I still had fair acceleration and I could cruise the motorway, say at seventy, with the engine just purring. If I had gone up three more teeth on the sprocket that would have made pulling away a struggle but the top end would be phenomenal. Well, for a Shovel.

Fuel: Paul had three choices in the fuel introduction he used.

- Nitrous. Paul feels although it would do the trick, it just wasn't cricket to use a gas to enhance the liquid. The Nitrous would ensure great acceleration up to the first timer then the turbo could take over for the acceleration to top speed. Nitrous doubles the revs in an engine very quickly, so using it from the start line makes a lot of sense. All the nonsense you see in films where they wait till the last minute is stupid. Why get the revs up to near max then try and double that? You'll just blow the engine.
- Turbo *and* supercharger. This would look great. Just imagine all the tubes needed to get this baby whirring. However, the weight gain would be on the obese side.
- Variable geometry turbo.

It would all need experimentation on a truly interesting level.

Then there are the fuel choices.

- Leaded, high octane (MON, not RON) petrol. The *real* stuff, not what we road users have to suffer with. Actually 135 MON.
- Methanol. Cooler burning so the intercooler could go.
- Nitro-methanol. The *nasty* stuff that really does the business.

Paul was hopefully going to use petrol and the variable geometry turbo. That was the initial plan.

18th September was the last showing of Svarog. This was purely for Airbus to have a look at what was going on. Paul didn't expect to get any more sponsors from this show. It was basically a goodwill move, Airbus had been good to him.

Paul picked me up to go and "chat" with Garry. He has many anecdotes about the build and the history of Svarog. On the way over Paul was talking about Garry. The man is a self confessed 'lazy' git. It was 2.30 pm and he had been out of bed for an hour or probably less. Hey, the guys not only disabled he's retired. But also, says Paul, he is the most generous person he knows. "Great." I thought. "I'm bound to get offered a cup of tea then." In Garry's garage, we chatted about his Shovelhead and its inaugural place in the scheme of things. While nattering Garry offers me an SU carb. As you may know, these are a beauty to behold, "Thanks," say I, "But I can't afford that at the moment and I have a few carbs on hand for emergencies as it is. "I wasn't selling it to you, I was giving it to you!" he said. That is generosity. Oh yeah, I got a cup of tea too. Bit by bit I was building a mental picture of these guys. A 3D picture.

Suddenly it's "Treacle Time" when events start to move really slowly. Paul was suffering from the anticlimax of the end of the show circuit time. The wiring needed sorting and he was mentally exhausted. Not being able to spend quality time on the bike was bringing him right down. Unfortunately, in that state anything that one tries to do ends up needing doing again. It's best to sit back and rest. Pushing just tires you more.

At this point, we decided to write the book first, make the movie, get the T-shirts and "kiss me quick" hats made and *then* go for the attempt. It's good to have everything planned out before you start. Nevertheless, things were coming together. The suppliers of sensors had been found, the wind tunnel was set up and the "Devils Spaghetti" was being planned to energise the bike. And I was coming off the beans and cabbage diet as there was no need to test the farting in bottles theory anymore.

The wind tunnel is set up for December. (Results in the previous chapter.) All hats doffed to Ian Dugary who had sent Paul a check from Aberdeen for a not inconsiderable amount. Top man.

Thoughts of buying proper welders, dyno machines, milling machines all filled Paul's head. He wants to be set up for any eventuality, but he must also be a little sensible.

Paul showed me pictures of the mount for the wind tunnel testing and how it holds the bike above the bed of the tunnel. It measures all the forces on the bike. One thought that came to mind was, "What about the effect that the ground has on the air turbulence?" There's one thing that I know about the Heisenberg Uncertainty Principle and that is, you can look at it all you like but you can never tell what the feck is really going on. That's science that is!

The intercooler was dropped off to go onto an ejector rig. WTF? Well, till very recently this was more top-secret kit, and used to test the aerodynamic qualities of missiles. This piece of kit doesn't blow, it sucks. Sounds like something to be tested? You'd have second thoughts when you hear it sucks at supersonic speeds. You don't want any of your tackle getting in the way of an air stream that moves in excess of seven hundred and fifty miles an hour. Or maybe you do, who am I to judge?

It looks like a twelve-foot canon. It's rigged up to a massive compressor that produces hundreds of pounds per square inch. Somehow, (See how technical we can get?) that gets blasted with very high pressure air then, through a series of pipes and valves, converts that into suction through a long barrel on the front. Ideal for a zombie uprising. Just funnel them into it and whamo.

All down the sides of the ducting that Paul had made, they are going to drill holes and fit an array of Pitot tubes that look like fibre optics. These measure the pressure at any given place. The intercooler is fitted inside this and a range of speeds up to three hundred miles per hour will be tested. That will measure the pressure differential on each side of the intercooler. At a certain speed, turbulence will build up and somewhere (you can find out a bit more in the wind tunnel experiment in the last chapter.) the movement of air through this will stop. Nothing will pass through and that is what they need to find. Then they will know what and where the maximum flow is as that is where the best cooling takes place. This sort of stuff is rarely done. Most tuning, as anyone who tunes will know, is not done this way. Usually it's a case of *"Let's try this.... Oh bugger, that didn't work! Let's try that.... Hmm, a bit better."* Etc.

Then, all of a sudden and with great gusto, everything stopped. There was nothing going on in the garage. Nothing happened anywhere in the near vicinity for about a month. But it was all about to kick off big time. The wind tunnel was all set up, the wiring had been sorted and all the sensors had magically found their way down to Paul's garage. Well, no. None of that.

The wind had been blowing, the snow had been falling and the temperature was not conducive to going to the garage. This was an ideal set-up for Paul to take a well-earned rest. He had been verging on exhaustion for ages what with work and the bike and the website and, well everything. Sometimes you just need to have a break. So with a month of down time in the bag, I thought it prudent for me to drop into the "Team Night" which happens every Thursday, wish the lads a Happy New year and just chin wag for a half hour or so. Nearly two hours later...

Even when nothing is happening something happens. Paul had found some nice wooden boxes in a skip, the new ceramic wheel bearings had turned up, four thousand pounds worth of sensors had been well and truly looked at on an American website and a shiny new centrifugal clutch was sitting upstairs waiting to be fitted. Paul explained how the sensors mingled with the engine management system would not only do away with the need for a dyno, but map everything out from the amount of air in the fuel to the pressure and heat of burnt gases with everything in between. With this kept on digital record for every run, a huge amount of data could be gathered in a short amount of time. This would lead to 'knowledge of what to do next' for every consecutive run.

The centrifugal clutch sounds like a scary piece of kit. Once the engine hits a certain RPM, the clutch engages and won't dis-engage until the revs drop below that point. This means, you guessed it, when you're rolling, you're rolling. Want to stop? Well, you just have to wait for that, and pray that some pigeon doesn't decide to land on the track in front of you. At least Paul has some nice wooden boxes.

Paul sent me a text saying he had something he wanted to show me. My interest was piqued. I was expecting some new bit of kit, or maybe something that has been in the pipeline for a while. After all, there is a lot of kit in other people's hands that Svarog desperately needs, but this was something completely new. It's funny when you have a "rolling stone" that it seems to actually gather moss as opposed to not, as the saying goes. With the stone being the whole project it's amazing what turns up unexpected.

Paul had heard of a new bit of software that Airbus had just purchased. In the aircraft industry a lot of testing happens before the finished article is allowed out of the oily grip of the mechanics. No one really wants to be at 35000 feet to find that the wings don't actually do the job that was intended for them. So the crew would build a wing, attach a lot of sensors and then put it through its paces. Testing of *everything* can go on for five years or more before anyone will put their money into the equation, sing out "that'll do!" and the plane rolls out of the hanger.

So these bright sparks have come up with the software that negates all the 'hit & miss' testing by doing it virtually. These boffins had a little ace up their sleeves as they were also in on the design of the original V-Rod and worked closely with Porsche and Harley Davidson to bring the project to fruition. (As an aside, the rumour that goes around about having to 'add' vibration to the engine so people would equate it with being a true blood Harley is true.) They do in fact supply this software to all the Big Boys out there: Mitsubishi, McLaren, BMW, in truth, everyone who does something BIG.

So all this talk in the canteen about this new software got Paul salivating. I'll let you in on the details of the software in a moment. So Paul goes and sees the 'clever' guys who are going to use this at Airbus. And yes, you have to be clever. His obvious question is, "Can I get access to this?" "Oh, no no no no no! A licence for this is £70,000+ and we only have four licences." So that was the end of Paul's trip to the IT department. Or was it? The very nice man in charge said he would e-mail the software developers. Paul left and went back to his desk. Within ten minutes, Paul gets a call. Can you believe it, the man from 'LMS Imagine Lab' wants to meet Paul and help out. He is going to let Paul have a licence for three years. WOOO HOOO! For the British record and the future of Svarog, it will save so much time, energy, disappointment, money and heartache. And they get their name on the side of the bike. Bargain!

Paul drops in to show me what £75,000 worth of software looks like. LMS have stuck to their word and delivered in the form of three CD's, and some booklets with lots of words that I have never come across before. We chatted about the potential from this small unassuming package. It is huge. Way bigger than either of us can imagine at this point in time.

The main part of the software will handle the totality of the bike. Then a sub-program called Virtual Lab covers the drive line, engine etc. It can be programmed right down to the weight of the pistons and the cotter pins. It's all in there. This does not leave out one iota of information, so nothing is left to chance. Even the durability of certain metals used in any part of the machine. I asked if and how it would factor in the road surface. A quick search showed rider interaction and other non-linear factors. It took in suspension and any modification needed for each simulation. It could factor in different tyre manufacturers and the effect each type of tyre would have on the rest of the bike, power-train simulation... the list went on. There probably is something in there about tarmac but it is such a huge piece of software to search through that it probably doesn't matter for now.

Paul has bought a laptop purely for Svarog and will be home loading this stuff before I can even finish typing about it. Broadband permitting.

I seem to have a picture of the skinny wiring loom after about half was removed. I'll put it here. I'll also put one on the website so you can enlarge it. Go to Rodlawless.org. It'll be there somewhere.

We chatted about where things were at, and the subject of the website arose. He felt that things of this ilk had not gone as planned. It was a case of too much too soon. He wanted to start again and asked if I knew of anyone who would fit his parameters for "Walking the Dog" in this particular arena. IE, someone who has a bit of knowledge about the project, a love of bikes in general, lives not more than two miles from him and knows how to turn a computer on. Although he didn't push it I knew what he wanted as he had alluded to it before and without too much resistance I offered my services. You know when you just get a feeling that sometimes the right thing to do is the thing in front of you? So I offered to 'work' his website for him. I have been asked on a few occasions to lend myself to people's websites and have always refused. With the Byzantine site, the vibe was right and I knew I would feel guilty if I didn't step up to the plate. As it happened I felt very positive about it.

It was at this point that Mrs L unceremoniously brought out some lunch for us. Paul got a bacon sarnie and I got the veggie alternative of a fried egg roll. As I took my first bite Paul thought, "I won't tell him, he's a big boy." The yolk oozed onto my palm and all over my leg. Apart from that, it was lovely.

So Paul left a happy man with plans to keep me busy on the website front. I shook his hand allowing him to partake of some of my yolk. He said it was nothing compared to what he was going to lay on me in the coming months.

Chaos Theory:

In every system known to man and, indeed, any that are not known, there comes a point in the growth of that system where chaos naturally gives way to order. It is called a bifurcation point. (Splitting.) It happens in the division of cells and the forming of stars. It shows itself in the growth of a human or a multi-billion dollar corporation. It is a point where all possibilities converge, where any of an infinite number of outcomes can transpire, yet only one does. It could be the following of any of a multitude of paths or the death of that system. It seemed Svarog had just gone through that stage. The wiring loom had been returned in a svelte condition and was poised for duty.

The sensors had all been fixed in their respective ports and Paul was excited. He reckoned that within a week Svarog would be back together with eighty percent of the ancillary items in place. By the end of the month, the baby would be breathing. Can you feel the buzz?

There was even a bunch of no less than seven magnets on the front wheel. If you've ever set up a push bike speedo on a custom build then you know how this works. That usually takes just the one magnet and one sensor. With seven, the sensor is working way faster and taking in much more data.

Paul now wants to get a good weather station. That way he can figure in other variables like the temperature of the air, humidity, barometric readings, pollen count and will he need a hat.

I caught up with Paul to go over re-vamping the web site. The project was getting to the place where a new page was being turned and a new website would be the order of the day. As we chatted, Paul mentioned that he had made a decision. Bonneville and all its glory, bad surface, dodgy weather and difficulty to get to, was not going to be on the agenda.
The British record was where Paul was intending to break all pre-conceived ideas and bring the sport home to this fair isle. After all, what have we really done over here in the past 50 years? The Americans have been plugging away at the World Record for years. Now and again someone goes out and does something over here and it is lost in the dust of old records. It has been ten years or more since the last British record was set. *(That all changed since starting this book.)* We need to get back in the game. Paul was going to change the rules and make history, Big Time.

Imagine a streamliner built for British conditions. Short and sweet, it would be a miniature fighter jet, but without the weapons. Shame.

Money is still a big issue and sometimes 'needs must'. Paul had taken on more work and now had two jobs running side by side. This left very little time for Svarog. To compound matters, he also found out that he had no way of logging all the data from the data logger. On many modern machines, the info can be stored and retrieved at a later date. Not so with the V-Rod. Seeing as Paul had completely negated the need for running on a dyno, he had to buy a small laptop that could be fixed to the bike to record the all important information on each consecutive run. Run your eye over Svarog and you will soon see that

Paul has a 'thing' for Velcro. Well, it is a very useful invention and holding the new laptop on the seat should be no problem at all. This could be interesting. I'm not saying I don't trust Velcro, it's just... well, yes, I am saying that actually. I'm glad it's not *my* laptop.

He estimated about thirty hours of work left to do until the first start up. Seeing as he has a few days holiday later in August that should see the beast breathing. But the breathing stage was still a long way off being ready for a test ride.

At last the big day arrives. It is actually big in more ways that one. Not only is it the day that Svarog breathes for the first time in three and a half years, it's also Paul's 56th birthday. Add to that the 5th anniversary of the bike landing on British shores from America and it all mounts up. It was also, I'm reliably informed, my wedding anniversary. Happy wossname, Mrs L.

"Coincidences are the Universe sending you messages." Deepak Chopra.

The last few weeks have been intense for Paul. Things that should take a certain amount of time took twice as long. *"Should be ready next week!"* kept creeping into the future. A furious spending bout sees funds from all the sponsorships dwindle till the kitty had lost its stash. But then a lot of essential paraphernalia has been bought. Some of it expensive and some of it frustrating. Petrol lines that should have fit had to be butchered. Things that could go wrong did, but it all lead to the perfect day to fire it up. Perfect except for the failing light and the rain.

With film crews from the BBC, photographers and journalists vying for pole position our own photographer, Mick Kirton, had to sit back for a while.

Around fifty people turned up on the night and braved the weather to hear... nothing. The battery was dead. This gave Paul a nice little test of his *"Hope for all but expect nothing"* philosophy. I think he expected the battery to work like it did for the film crew only two hours earlier. The trouble was that when Ethon fetched the battery, a spanner that had been left on top touched the terminals and knocked out what little there was left of that ageing lead lump. Once jump leads were brought into service, the crackle of Svarog drinking down that MoN 130 petrol filled the air. And yes, you *can* call it *'petrol'* as opposed to just fuel.

The upshot of the evening made itself known over the next few days. Not only were we in the local Evening Post but the Mail and the Sun also featured an article. On their websites you can actually see people's reactions to the project. They range from "Good on you mate." to "Well, I can go fast on my bike, so there." And the obvious, "That's impossible."

As the ancient Chinese proverb proclaims
**"The person who says it cannot be done
should not interrupt the person doing it."**

I popped round on the next Sunday to find Paul getting to grips with the intricate and sometimes complicated software and sensors. It wasn't obvious why one sensor would withhold information on one cylinder but when swapped for its counterpart on the other cylinder it all started singing and dancing. But beyond small niggles the sensors were doing their part, the computer program was doing its bit and Svarog looked like he could do with a little tweaking to iron out some rough spots.

About a dozen times, Paul fired him up and rattled the windows of persons close enough to be affected. One such neighbour was a mother and her son who walked past. The little boy was wild about the bike. He had the newspaper cutting in his pocket and just loved the whole idea. Paul and Svarog were on their way to being local heroes.

But it was going to be a slow old climb. The Power Commander that came with the bike wouldn't recognise the new interface and problems were arising. There was only one thing for it, he had to buy a new Power Commander. Estimates flew around the garage of close to a thousand pounds. At it happened he found just the one he needed for five hundred quid. So that was five hundred saved. If that sounds like female shopping philosophy, it's because it is. Just when he thought all the major expense was over with, this sprung up and bit Paul on the arse. All the same, there were a few benefits. The module was better in a few concise ways that would enhance the operations of the fuel injection. The old module would only read the atmospheric pressure whereas the new one would read the fuel pump pressure. All those wild spikes on the software graphs would even out. Hopefully.

The end of the year had sprung unbidden like a coiled kitten up the trouser leg of time. It was too late to even think about getting onto a track now. But a dyno could do a lot of damage to the lack of data that we needed for Mitsubishi. Also, Airbus were taking notice of the renewed media interest and Paul had managed to get them to agree a day in the wind tunnel. This is where the LMS software would really pay dividends. Once the data from the wind tunnel was factored into the equation, Paul could run simulations on his computer from that point on. There would be no need to go back to the tunnel to gain a better perspective of any new bodywork.

Paul brought the new nose cone that Andrija had designed, around to me. He had been trying to drill it to fit the plates on the front forks. Although he was very careful and went up in one-millimetre increments, very slowly, it still cracked the moulding. This thing has been built using a 3D printer. You know the sort, got them at Argos, cheaper to buy a new one that replace the ink. Um, no! This is the one that fires two or three lasers into a field of suspended plastic particle dust. You can pick up a tub of this dust for around £6,000. Give or take a coronary embolism. They used nearly a whole tub to build the beast.

To make it a bit more stable for the wind tunnel test, Paul asked me to fibre-glass the inside. After all, a two hundred mile an hour wind can do a little damage. We didn't want bits of the thing flying off into the multi-million pound facility. They wouldn't invite us back if that happened.

You can just about see the crack in this photo. It runs just over half way round under the midsection at the front.

And I can just imagine a bloke in a white lab coat running over to the tea trolley to remove the sticky buns. Our popularity would be lower than a suicide bombers self-esteem.

I made up a batch and slapped on a thick blob of goop. At the exact same time, it hit me that the two might react. Ooer! Plastic and resin? Svarog (The god) must have been smiling on us that day. Paul went home happy and I breathed out.

So, once it has been tested then a proper mould will be taken, and carbon fibre utilised to build a lightweight and very robust nose for the bike. This will not look like your average drag strip hooligan, and it won't go like one either.

The years teach much which the days never know.--Ralph Waldo Emerson

Yet another team meeting. Unbeknownst to me, Paul had in his possession a piece of machinery that plucked the "WOW" factor right out of me. Mitsubishi had come through with the most phenomenal looking piece of kit. Being all in pieces Paul could show me the business part. The heart of the beast as it were. As Garry pointed out, "How do you go about making something like that?" (This picture is the near finished article as I didn't get a photo of all the bits. Sorry.)

As a youngster, I had made a rather unsettling discovery; The whole universe is governed by mathematics. For someone who saw the use of numbers in different configurations as a nonsensical waste of time, I could have been out building go-carts and stuff, this revelation was daunting. Remembering formulae was beyond my capacity. Was I ever going to understand turbos let alone my life? Quick answer, no!
So, if Paul's and Mitsubishi's numbers did whatever intimidating nonsense it is that they do then this turbo was going to be the lynch-pin in the performance of this particular V-twin.

Rich hadn't been idle either. He had been slowly drizzling himself over the wind tunnel data, (He is more at home with these 'number' things than anyone I know.) and had come up with a three-dimensional graph that showed how much air was going through the intercooler and at what speeds. For those who find this sort of science interesting, at 100 mph the air in the front of the intake is doing 75 mph and at the back, it is doing 68 mph. At 150 the air in the front is doing 146 mph and at 200 mph the intake air is doing 192 mph. The faster the bike goes the faster the ratio of air in the intake. Not what any of us had really expected as it would have seemed that the more air you squeeze into that tight space, the slower it would want to go. Andreas had done a superlative job on that design. In fact, on all the design fronts that he has had a hand in, great strides have been made. The tail fin would definitely do the job of keeping the bike in a straight line and the nose cone, if I can call it that, was going to make the 'going' easy. Well, easier.

Paul had bought himself a little toy in the guise of a digital Tig welder, and was happily destroying small bits of metal in the search for the ultimate weld pattern. He was soon going to be having a party with the other bits he had made, the flange, an inlet cowl type of a thing that looks really funky and of course all the bits of tube etc. for the rest of the exhaust system. Once made up Svarog would be fired up and a lucky piece of tarmac somewhere around the country was going to have its pants blown off.

I turn up late to the next Team Meeting. In attendance Paul and Rich. The air is blue as Paul gets animated about something Rich has said about wind resistance. Sometimes it's good to blow off steam. I sit quietly watching as the blast furnace that is Paul's passion for this project lets the pressure out. Like a good storm brings fresh ions this tirade ended with an uplifting atmosphere. They were soon hugging and smooching and doing that thing where you link arms and drink tea simultaneously. It's good to see the extremes coming to the fore. We are all human after all. Yes, even Richard.

Down in the garage, the turbo is all fitted, but some wiring still needs to be done. Plus, all the data from the wind tunnel has been transformed into some new parts. More aerodynamic fairing bits have been constructed with the 3D printer. Now everything that was once out in the open will be behind some sort of screen. The overall surface area has grown but the drag co-efficient has been reduced. I think this is what had started the... 'discussion'.
On the down side, the old LSR had been broken at the Straightliners. Another seven MPH

had been added taking the new record to 229. It was a tricked out Hayabusa that has caused the hiccup in the proceedings. But no matter, Paul will still aim above that.

One of the last things we covered was the clutch which has an internal micro switch. This allows a set point for the clutch to engage. You hold it wide open and the engine will only rev to, say, 3,500 revs. Pop the switch and it lets out the clutch at the same time letting the revs rip up the scale. Marvellous.

Then... Once more, not much for over two years. Nothing seemed to be happening. I just sat back and waited as I had been wrong in that assumption before. Like having to endure Religious Education when I should have been learning something useful, life seemed to get in the way. Time once more drifted on in its relentless way. I was on the verge of setting up a TV crew through my cousin's involvement with Sky TV, but thought I would wait till something moved.

The team still met on a Thursday but I had a feeling something was out of kilter. We were close. Very close. But there are things that drift unseen in the human psyche. We could see that Paul needed a rest. But what we couldn't see was what *he* also couldn't see. Something malevolent. But that will have to wait for later, now I want to talk about....

Chapter Eight

TURBO'S.
Power to the People.

"There is no great genius
without some touch of
madness." Seneca

The way up and the way down are one and the same. Heraclitus.

You may be looking at the diagram and thinking it's self explanatory. Then again you may not, so read on. You are not as lucky as me, I had Paul to tell me. You, on the other hand, only have me to tell you.

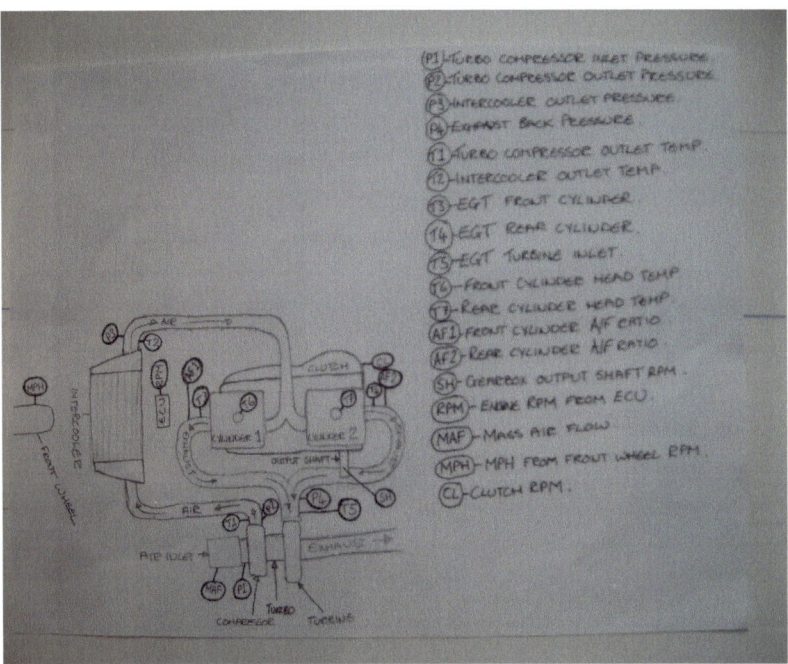

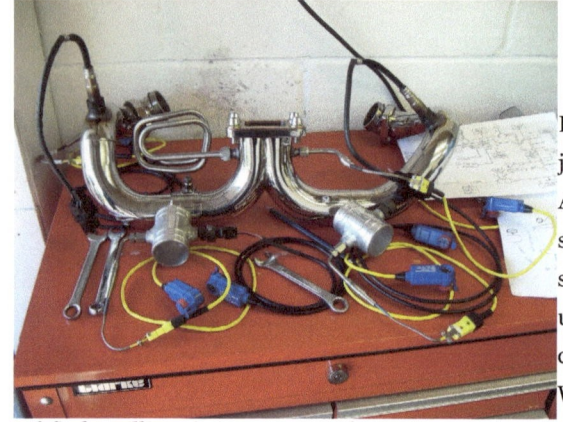

I learnt a lot about turbo design just from having Paul go over it. As they say, if you want to learn something, then explain it to someone else. Or is that "If you understand something then you can explain it to others" Whatever. I'll wait while you go and find a willing victim, I mean subject, so *you* can explain it once I'm done. That way we all learn something.

Okay, back now? I'll carry on. Just imagine you are a ten-year old and that is what I'll do to get the message across. That way I can talk down to you as if I really know what I'm going on about. Whereas you can act all impressed that I know it.

Now bear in mind that Paul thought he knew most of this stuff until he looked deeper into it. His education in all things fast spans at least thirty years. His new education on measuring "all things fast" flew into him like a wasp in the crash helmet. It makes you wonder how you didn't see it coming. And when it's right there in your ear you need a way of getting it out. (So others can see it.)

This set up will allow so much data to be collected and collated. Against RPM you can map out cylinder head temperature and fuel/air ratio etc. Instead of just guessing and then taking it to the dyno to find out if your guesses were correct. And, as you may have figured, this will save not only time but breakages.

(Picture shows original turbo and placement.)

All the sensors run into the Data Acquisition Module via a single loom of wires run in sequence. This keeps the whole show down to a minimum. This logger has a vena input, analogue input and RPM input and can monitor 67 channels.

The great thing about it is that you can take out the little memory card and store it or retrieve all the data. Or you can plug it into a PC and read it in 'real time' and adjust as necessary. The digital age makes things so much easier.

It's nice and small, has fifty seven channels and sits under the seat on the tail piece. Thankfully.

I intend to explain this as if I were the one listening. If you have way more knowledge then great, but there are people like me out there. Where to start? I guess where the air goes in because that starts the whole "Turbo" process. It's quite simple.

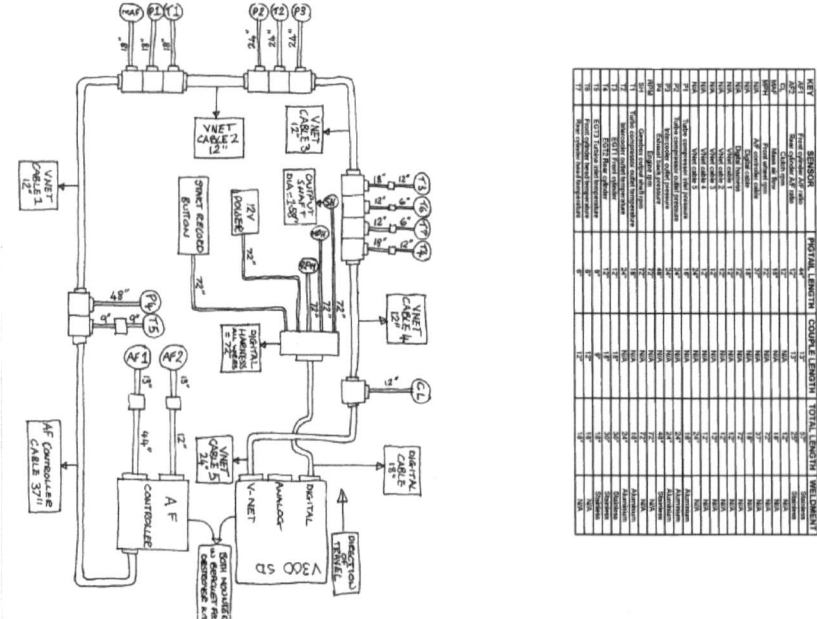

P1: <u>Turbo compressor inlet pressure</u>. External to the inlet, air pressure is 14.7 psi. At the point where the compressor is spinning at 120,000 rpm, (Yes, you read that right.) The pressure drops. So that has to be measured.

The compressor is like a pump but for air. An air pump, if you will.

P2: <u>Turbo compressor outlet pressure.</u> Thep pressure of air that the compressor wheel makes. The Boost. Told you... Simple!

T1: <u>Turbo compressor temperature.</u> As air get compressed it gets heated considerably. Here's a molecule lesson. If you pack molecules together, even though they don't touch, they cause heat. It's the same as filling a room full of twenty-one-year-old Puerto Rican girls and you sit in the middle as the nucleus. The temperature rises.
So to understand the efficiency of the compressor wheel, the size needed for a particular RPM range and how much pressure and what temperature it adds to the combustion cycle, T1 and P2 work in unison. The hotter the air the less dense it becomes and therefore the

less power the engine makes. Plus with a temperature on the piston surface around 200 deg C before any interference any further heating needs to be kept to a minimum. Without an intercooler, the engine is sitting on the edge of being a silver coloured chocolate fountain.

Then the pressure and temperature of the air on the way out of the intercooler are measured by P3 <u>Intercooler outlet pressure</u> and T2 <u>Intercooler outlet temperature.</u> This gives the integral efficiency of the turbo. Paul was looking to get 70 to 80 % efficiency out of his particular set up.

That's where the business gets done as the air goes right into the engine at that point. All this has to not only be done with the most efficient set up possible but also the equipment must be kept as small as possible. This is why we were so glad Mitsubishi were on board. They know their onions.

T6 and T7 measure the cylinder head temperature and are a ring that the plug goes through. Thermocoupler sensors which sit right at the hottest part of the cylinder. You're still thinking about the Puerto Rican girls, aren't you?

T3 and T4 are <u>Exhaust gas temperature</u> <u>sensors.</u> Does what it says on the tin.

AF1, AF2: Reads the residual oxygen after the burn and the software calculates the exact air/fuel ratio at any given time. Now we are in the clever stuff department. Not only does it measure that but it doesn't melt from the heat.

Following on with more clever stuff, P4 and T5 combined give exhaust pressure and gas temperature. From that, it is possible to work out what sort of cam-shaft is needed and the size of the turbine and its aperture.

Non-turbo sensors include SH which is the gearbox output shaft and CL. The latter measures the revs of the clutch. Mixing the two tells us how much slip there is.

The nifty thing about the clutch is its centrifugal action. It doesn't slip much if at all but on the other hand, once you are going then you are *GOING*.

With the MPH and RPM of the engine, there is, all in all, a lot of information. Paul can use most of it and Mitsubishi use all of it. So the data logger with it's fifty-seven channels can take *all* this information and store it on a card. Then it can spew it out onto a laptop.

If you are racing a turbo engine this is all the information that you need. It can all be put on a graph and read by someone with a higher IQ than me. Then that can be translated to the setup and viola... fast.

Paul reckons there is a market for turbos, not only for V-Rods but for your average Big Twin. Even for the old Shovels. It is something I started thinking about for my old '78. But 'thinking' and doing are two different things so maybe there is a market, but who is going to make it happen? I guess the buyers if they are also doers.

Chapter Nine

Venues For Speed Records.

(Alphabetical order.)

Sat amidst the northern Nevada section of the Great Basin, a dry remnant of the Pleistocene age is the **Black Rock Desert**. Once, this place was a trail for early emigrants doing about three or four miles an hour on their way to California. More recent times has seen Mach 1.02 across the same landscape. But then, rockets are a tad on the slippery side when up against cows.

If you should drop in at the right time you could also witness the "Burning Man" accumulation of hippy types doing their thing.

Sandwiched nicely 'twixt the M3 and the M25 south of the metropolis of London, is the little sun-drenched town of Weybridge. As well as being one of Britain's first airfields, it was also the world's first purpose-built motorsport venue, **Brooklands**. Opening in 1907 it worked till 1939. However, time and the 'tide of change' stand still for no man and no circuit, so now it is the home of the Brooklands Museum and a venue for vintage car, motorcycle and other transport-related events.

As an aside, the author is proud to say that his granddad raced there in its heyday, in his home built cars. He even came home with silverware now and again, which was an added bonus.

The most famous venue by far is, of course, **Bonneville Speedway.** Although the Salt Flats are vast only a small part is marked out for motorsports. It was first used in 1912 but didn't become popular till the thirties when the likes of Sir Malcolm Campbell peered off at the horizon and wondered how long it would take him to get there.

A great part of it is navigable but there are parts where you could lose your oxen and cart. That happened a lot in the early years of its discovery. So, take care if you wander from the beaten track.

Head West-South-West for a couple of thousand miles or more, and you could find yourself near **Daytona Beach Road Course,** with 4.2 miles or straight tarmac. Instrumental in the formation of NASCAR, (National Association of Stock Car Auto Racing), it originally became famous as the location where fifteen World Land Speed Records were set. Now it is famous for muggings on the beach and tassels on your chaps.

RAF Elvington was an airfield located in Yorkshire, that operated from the beginning of World War Two until 1992.

On 3 October 1970 Tony Densham, driving the Ford-powered "Commuter" dragster, set the official outright wheel driven record at Elvington, averaging 207.6 mph. over the Flying Kilometre course.

Others, Like Henry Segrave (Who had been setting LSR since 1926,), Colin Fallows and Richard Hammond have all had a plug at a title there.

Taking a leap in the alphabetical progression stakes of this section, we land squarely on **Jabbeke.** In the Belgian province of West Flanders sits the municipality of five towns, one of which is Jabbeke. Being Belgium it is mostly boring so no photo and we now skip to;

Australia. Much more exciting is **Lake Eyre.** Pronounce it as you will although apparently it's supposed to be aer. See. Doesn't help, does it? Being the lowest point in the continent it sometimes fills with water, then it is the largest lake in Australia. Although situated in the colder, (Than the north, not the UK.) South Australia it is still smack in the middle of a desert, so the water thing is not such a problem for racing.

Coming back half way round the world to a less inhospitable environment we happen upon **Pendine Sands.** The seven-mile long beach on the shores of Carmarthen Bay in Wales is not only easier to get to but has more tea shops than Eyre. That's always a selling point in my mind. The early 1900's introduced the sands to the petrol engine, and in 1922 the Welsh TT motorcycle event got under way. Shortly after, in September 1924 Sir Malcolm Campbell set a World Land Speed Record of 146.16 mph in his Sunbeam 350HP car, Blue Bird.

Motor Cycle magazine described the sands as "The finest natural speedway imaginable". Obviously then it was straighter and smoother than any major road at the time. Funny how some things don't change.

Four more record-breaking runs were made on Pendine Sands between 1924 and 1927; two

more by Campbell, and two by Welshman J.G. Parry-Thomas in his car Babs. The 150 mph barrier was finally broken, and Campbell then raised the record to 174.22 mph (280.38 km/h) in February 1927 with his second Bluebird.

(Photo; Campbell at Pendine in 1927 but not in Bluebird.)

Sadly, disaster visited the South West Welsh coast. On March 3, 1927, Parry-Thomas attempted to beat Campbell's record. On his final run hitting around 170 mph (280 km/h) the exposed drive chain broke and partially decapitated him, Out of control Babs rolled. Parry-Thomas was the first driver to be killed during a world land speed record attempt. This marked the end of an era and was the final world attempt made at Pendine Sands.

RAF Fairford.

Big, flat and host to many large shows and some with old bikes. The US Air-force only moved out in 2014, so I can't tell you very much about the place. Hush hush and all that.

Southport (pronounced sa℧θpↄrt.) There we go again with the pronouncing. As you may guess if you didn't know, Southport is by the sea, in Merseyside, England to be precise. Why a northern town has South in its title is beyond my meagre comprehension, especially as there are no ports to the north of it.

Placed neatly on the Irish Sea it is bordered to the North by the wonderfully named Ribble estuary and sixteen miles to the south we find the also wonderfully named Liverpool. No swimming for vegetarians. Not only does Southport have the second longest pier in Britain, it also has one of those beaches that people like to go fast on, especially in the mid-1920's. 2016 saw a re-enactment with Sir Henry Segrave's Sunbeam Tiger rolling down the sands again for the ninetieth anniversary of the record.

Verneukpan (Not even going to try. Pronounce it as you like.) is a wide dry salt pan south of Kenhardt, between Swartkop and Diemansput in the Northern Cape, South Africa. *Verneuk* is Afrikaans for 'to trick, mislead, screw or swindle'. Maybe that was true for Malcolm Campbell as he was screwed out of a record by the harsh surface that ate his tyres. Perhaps if he had taken note of the bunch of other misfortunes that plagued him, from losing his briefcase to his plane crashing into a tree, he may have just gone home. Others haven't faired so well there, and in 2006 on 27th June Johan Jacobs died in an attempt doing over three hundred miles per hour.

It may be flat, but it isn't a stroll in the park.

Chapter Ten

The Attempt.

"Find the middle way."
Taoist Aphorism by Lao Tsu.
Tao Te Ching

Let him that would move the world first move himself.
Socrates

So this part of the book hasn't happened. At least not at the time of writing. But from the things we call 'bad' come things we call 'good,' and someone has bought Svarog with the intent of racing. I will endeavour to follow the progress and hope an article can find it's way into a magazine one day.

Also, just before Paul sold it, there was a very interesting interlude. Someone who had been looking to buy the bike dropped in to see Paul. He is an assistant director and associate producer from Hollywood. Working on films like Die Hard With a Vengeance and that sort of thing for many years, he wanted to buy the bike with the intent of making a film.

He didn't end up doing that but the good thing is I got a great idea for a film script and I shall be working on that soon.

Arse About Face

There didn't seem much point in writing a preface after all the real 'meat' is in the book itself. I don't know about you, but I very rarely read much more than the first paragraph of a preface. We all know what going fast means, either from first hand or maybe the *"knowledgeable person"* in the pub has told us.

'Arse about face' seems more pertinent. Especially as every ending is the beginning of something else. The future, folks, that's where this is heading. Time now to face *your* future. What drives you, what dreams do you have? What do *you* want to achieve? It doesn't matter what it is, it could be to beat the record. Maybe you want to build a great custom. It could be to earn five thousand pounds a year more than last year. Maybe it is to grow a better lawn or beard.

This story, so far, has been about Paul and his dedicated team. But on a deeper level, it's about you too. This story has spiralled out from the hub and touched more people than we can measure. If you are reading this then you are affected by it in one way or another. Maybe you had a little chuckle. Maybe you came up with a better idea. It doesn't matter, you are linked. Is there a book about you waiting to be written? Only you can make it happen. It may take some education of one sort or another. After all, if you're not growing (learning) you're dying.

Do you have ten thousand ways of not making a light-bulb? In other words, can you keep going in seemingly adverse situations? As the great Zig Ziglar once said; *"Your attitude determines your altitude"*. If you can cultivate a winning, glass half full, of better still *'all full'* attitude, you can achieve anything you want to. Pull out a piece of paper and write down exactly what *you* want to do. One life, live it.

Well, folks, that's the sum total of what we know so far. What we don't know is the final output in horse-power or torque of Svarog. We don't know what the top speed would be or how the acceleration would factor on a rider. That is not a lot in the grand scheme of things, really. In the construction industry, they expect a wastage of around 10%. I think we have beaten that.

"Your time is limited, don't waste it living someone else's life. Don't be trapped by dogma, which is living the result of other people's thinking. Don't let the noise of other's opinion drown your own inner voice. And most important, have the courage to follow your heart and intuition, they somehow already know what you truly want to become. Everything else is secondary." - Steve Jobs

Interesting web sites:

Go here for some footage of Svarog being fired up for the first time, and in the wind tunnel. Also there are photo's that didn't make it into the book.

Www.rodlawless.org

Needed horsepower;
http://purplesagetradingpost.com/sumner/bvillecar/bville-spreadsheet-index.html#Horse%20Power%20Needed

http://robrobinette.com/top_speed.htm

Land Speed Racing;
http://www.landracing.com/forum/index.php

Centre of gravity;
http://www.grc.nasa.gov/WWW/K-12/airplane/cg.html

Aerodynamic centre;
http://wright.nasa.gov/airplane/ac.html

3D design and test software
http://www.lmsintl.com/

The future belongs to those who believe in the beauty of their dreams.
Eleanor Roosevelt

Epilogue.

I guess you've waited long enough. Paul told me in the hardware isle in Tesco that he was just getting over a long illness and had given up on the Land Speed Record Attempt. That is the main reason the last chapter remains empty, I wasn't getting lazy with the typing. He just couldn't get the support he needed. The work on Svarog could take another year with funding of at least ten thousand pounds going into it, and he just could not do it anymore. Without more support, he would probably need another three to five years and very deep pockets. This takes a ton and a half of commitment. Once your commitment bucket is empty there is very little you can do.

I had been so inspired by Paul's project and thought I had filled my niche, to put it politely. I knew it was what I was supposed to be documenting. So I got down to it and even felt massive amounts of that somewhat fleeting human tendency, inspiration. I was on a roll, as were the whole team.

I don't intend to sugar-coat any of what I am going to tell you, that would do a great disservice to Paul and his team. Because this part of the story also has to be told and I want to tell it my way.

A few weeks later Paul gave me the deeper aspect to his tale. The general message is like this:
That grand era slotted between the sixties and the eighties is long gone never to be lived again. It was a time of freedom that wasn't quite realised in the sixties due to the legacy of the fifties. The seventies was an explosion of discovery and new horizons. No Big Brother, great birth control, advancing technology without the hindrance of computers. For me, I didn't realise how good it was. Looking back it was the best of times. For Paul, it was free reign to be who he really was. That may sound a little 'New Age' but when 'who he was' got up in the middle of the night and fucked off, Paul was left in totally another place, a region he didn't know. However, he didn't feel lost, he felt lighter. Although he is in no way happy about the situation as it stands as he feels he has let an awful lot of people down.
He had been doing quite a high powered job with the inherent stress related to it. He was running a side business to try and recoup some of the money poured into Svarog over the years and repay some debts. And he was spending every other waking moment on the bike, something had to give.

I've always been fascinated by mechanical movement. Why I don't know. Maybe it was Caracticus Potts in Chitty Chitty Bang Bang. If I follow the thread I see it all goes back to creativeness and inventiveness. Ultimately it takes me back to that other interest I have, the workings of the human mind. Unarguably the greatest bit of kit in the known universe. But the mind is not without its frailties, they number almost as much as its secrets. It really doesn't take that much for the mind to break down. We think of the body as being a delicate thing that even a minor motorcycle accident can ruin. But the mind, well, we don't even know what it is made of. Look how quickly we can change it. It seems to be just an illusion, albeit a persistent one. And then, of course, there is the brain.

Paul is not alone in this mishap of the mind. I have met many people over the years who went through similar things. Some cope well, others not so well. Normal everyday folks who for one reason or another have what is termed a mental breakdown and start listening to Leonard Cohen. A loose spark plug, a blocked fuel line, anything apparently minor can cause it. If we think about it, we might have all gone through a *minor* event of this type. We are pushed to the limit, by a person or circumstance, and something clicks inside. We say, "I'm not going to take it anymore." That, in a somewhat smaller nutshell, is what may well have happened to Paul.

Most people have a coping system. A crutch that helps them keep away from that sort of occurrence. Gambling, addiction, food, sex, TV, all sorts of things. Nevertheless, sometimes these things don't do what we want them to do, and the inevitable crack in the headstock appears. If seen we can squeeze some resin in and paint over it, but that just puts off the inevitable. There is no course or handbook that warns of these types of things. There is no real viable non-self-destructive strategy that we in the West have in place and so we succumb. Paul's wiring had burnt out, the person he was had gone. His direction was lost. I tell you, people, this is not a 'thing' we can see coming. It is just as much an accident as any other physical occurrence. Paul had done a complete 180-degree turn and now he couldn't face anything that had filled his life before.

Although there is still a slight stigma connected to a mental breakdown it is getting more recognition as a viable problem. It is also getting more prevalent in the 21st C., Strangely enough, it is rife in the NHS. As a society that is supposed to have been engulfed with labour-saving gadgets and appliances and virtually run by technology, we are under more pressure than ever. There is one theory that states the rise of communication availability corresponds to the rise in stress-related disease.

Watching BBC's Speed Dreams program about Bonneville reminded me about Chris Ireland's Breakdown. At the time he was writing for BSH, I remember it well as I thought him incredibly brave telling the world about it and hoping it would be a warning for others.

It has been well documented that a trauma or overload of stress can cause a re-wiring of the brain in such a manner as to render the thought process immobile in certain ways. What was once easy to cope with becomes a nightmare. Post-traumatic Stress Disorder is recognised as a debilitating disease. Not that Paul suffered from that as there was no life-shattering shock involved. It was a steady build up of pressure from all sides and the symptoms are quite different, except for one. Paul is not the person he once recognised himself to be. Some neural pathway in the brain has changed and the original mind configuration has gone with it. Synapses literally burn out.

Neuroplasticity is something that has only been discovered at the end of the last century. They discovered that learning something changes the connections between neurons and forges new pathways. New actions or thoughts turn on or off Genes within the cells which then make proteins which change the structure of the neurons. In a learning scenario, it can double connections in a matter of hours.

Neuroplasticity, at least the science of it, is utilised in recovery from anxiety attacks and as a natural activity in the learning process. What Paul experienced was a reversal of the operation. He lost all interest in the way his life had been going. It wasn't there anymore. This is not a conscious decision on his part. It is a burning out of the neural pathways that once kept that part of his life alive. He had been living with twenty-four volts travelling through a twelve-volt wire.

It has been found that what you choose to think about or imagine can actually change the structure of the brain. Change the neural paths. Paul experienced this in reverse. The probabilities for this new science are beyond miraculous. But for Paul, he has had to face the opposite. Yet up to that point, he had achieved so much.

A look at achievement.

I am of the belief that we are in this world to make some sort of change or contribution to it. We exist to change the world we live in. Which, inevitably, we will do one way or another. Personally, I have always strived to 'customise' my environment, build it the way I want it. In essence that needs some sort of success protocol in order to accomplish anything.

A success mindset, (As taught by millionaires the world over,) is one that takes *full* responsibility for its actions. It knows what it wants and 'knows' (not believes) it can have it. There is no room for doubt which means that the word 'belief' has to be dropped as it's counterpart is disbelief. If the mind entertains anything less that 100% then it fails. This is why we sometimes get the things we want and sometimes we don't. So to have belief in something, there must be an element of disbelief. Even 99.999% is not enough.

Hope.

This used to be touted by the good old religious people as something to strive for, (Along with faith and charity. Charity is the only one I accept,) and I'm sure they still do. Looking at the state of 'hope' it is obvious to see that it leaves one with an open attribute that doesn't garner moving forward. Having hope is like saying 'I wish' or 'let's see what happens'. It is not living in certainty. If you want to achieve anything, certainty is what you need.

That, in essence, is it. By the way, there aren't many success coaches and guru's teaching this. Most teach that belief is good enough as they take their cue from 'Think and Grow Rich' by Napoleon Hill. They are wrong, as ninety-five percent of their followers will attest. But that was a book written at the beginning of the Human Potential Movement, and is allowed to harbour some misconceptions for that fact alone. We have evolved in our thinking since that time. Read it book and replace the word 'believe' with 'know'.

The success coaches chuck in other useful tools such as writing out your goals to help clarify and crystallise them. Visualise or imagine that the goal is yours as if it already were. Do something in the physical world that leads you towards your goal. Feel excitement about the goal yet be detached as to the outcome. This is a bit tricky, but possible all the same. It is achievable if you have control over the greatest tool available to mankind, the mind.

This 'success mindset' is in play all the time. If we want to make a cup of tea we have to use all the above in order to get to the finished product. The only thing is, most of it is on auto-pilot and therefore not seen as an actual step in the process. Like driving/riding, it is done mostly by the subconscious, as we have learnt and incorporated the basics.

An Exercise in Success.

Let us make a cup of tea now, if only in imagination.

Do we know what a cup of tea is? (Can we picture or imagine it?)

Do we know we can have one? (Is it in the realm of possibility?)

Do we take responsibility for our actions to create it? (Am I actually going to get my arse out of the chair and head for the kitchen with that intent?)

Of course, that doesn't necessarily guarantee success in the moment. We might be out of tea.

There are a few of these teachers that say that we create whatever we get in our life by the decisions we make. The Buddha once said, "Our thoughts make the world." The more I look at that the more I believe it is true on many levels. Life is not just a random event that 'happens' with no input from us.

For a start, it is the action of 'thought' that designs everything we make. Our cities are proof of this, and so is the countryside if it comes to that. Look anywhere and man's influence is rife. In quantum physics it is said that our consciousness creates what we see. When a scientist looks for a certain outcome he gets the effects he is expecting, at least that is true with light. (The double slit experiment.) Everything is apparently made of light.

So what does this have to do with Paul and Svarog? Many people's inclination would be to believe that Paul did what 98% of us in the modern world do, cave in to failure thinking. (Although that is not actually the case.) There is no blame or judgement in that. After all, he had been working towards this for well over ten years. Who wouldn't have done what he did? I know I certainly would and probably a lot sooner. In fact I have on just about every project I have started. With some circuit burning out in the brain, the one that would have allowed him to continue his life the way he had before, then the result is inevitable. But then that is what we all do. All except the 2% of the population who are usually multi-

millionaires, billionaires and people who drink a lot of tea.

I still felt this story had to be told. There was so much effort by so many people put into Svarog, that something had to be kept alive in the project. I talked of responsibility earlier, and there is an ancient Hawaiian practice called Ho Oponopono. This basically uses four phrases to cure and help people. It is a clearing and healing technique that was held in secret for hundreds of years by Hawaiian Royalty. Read the book 'Zero Limits' By Dr. Joe Vittale to find out the miracles that have occurred. Anyway, Ho Opononpono goes by the fact that if someone comes into your life they are your responsibility. If they are ill, it is you who is partly the cause. This is not taking 100% responsibility it is 200%. It goes by the maxim that the world we experience is one that is co-created through thought by the people who interact with each other. In scientific circles, this is talked of as 'The Holographic Universe' and has very far-reaching aspects.

Ho Opononpono uses four phrases repeated whilst focusing attention on a particular person or situation. As we all know, words have vibration, even thoughts have vibration due to the electrical nature of them in the brain. Vibrations that are constant and repeated have an effect not only on the brain, but on other vibrations in the vicinity. The four phrases are "I love you. Thank you. I'm sorry. Please forgive me." They can be said in any order. Personally I have found just repeating one of the first two phrases is enough to make a difference. I use "I love you." And "Thank you." Try it on a long or boring journey when there is nothing to distract you, then you might find out why the ancient Hawaiian Kings didn't share this secret.

Holography in science.

Quantum physicists, amongst many others, explain the universe in mathematical formulae. The use of what is known as sacred geometry also helps explain how things from crystals and seashells to solar systems and galaxies, are formed. There is even a computer game now, No Man's Sky, that uses mathematics to create the world within the game. This allows the game to be infinite. If you go to one place in the game you may find a mountain. Go away and come back later and that same mountain is there. Okay, so there's nothing unusual there, but while you are not near the mountain, it ceases to exist. This allows an infinite game and not one bound by what can be fitted onto a DVD. Some scientists are saying real life works the same way. If no one has any conscious input on a part of the world it ceases to exist but is held in a mathematical temporal order. But since it is caused by the maths phenomenon that creates all the universe it reappears when consciousness is focused on it. (If a tree falls in the forest and no one is there to hear it did it make a sound? Was it there in the first place?)

So, are we in a hologram? Ancient Greek philosophy would say yes. "As above, so below" being the maxim of Hermes Trismegistus. God made man in his own image, the Bible says. These credos point to 'each part is a reflection of the whole' or holographic in nature. Are our scientists just finding out what has been known in many ancient traditions for millennia? I think so. We in the West are finally catching up. There is nothing new under the Sun.

If this hologram is being created by our own awareness of it, then it should be possible to alter it with that same consciousness. Can we create the environment we want? By thinking differently, it seems it may be possible.

Although the holographic model has yet to be accepted by the majority, (Most breakthrough scientists since Copernicus know what that is like) it does explain a lot of things that were unexplainable by the scientific community. As we now know, the Newtonian model that has ruled for so long is very limited. Michel Talbot's book, The Holographic Universe will explain much more.

In this instance, I take it upon myself to be partly the cause of Svarog not breaking the record. After all, I have known Paul and the team for five + years, and having anything to do with a venture will have some sort of impact on it. Whether any of this is true or not is beside the point. I still feel it is my responsibility to do something about the situation. You have my part of the story in your hands right now. Thank you.

What do we need to do to successfully complete a project?

First, we must know what it is we want to achieve. This cannot be some whimsical wish or ill-defined want. It has to be something that really hits the button. There is only one good way to find out exactly what is needed at this point, and that is to sit in silence and contemplate. Once we are clear on the topic we must write it out as fully as possible. This helps to cement the idea in reality. It brings it out of the realm of thought and starts to form it in the three-dimensional world. The process will bring to light things that we may not have considered. Plus it will give you time and space to think if it really is what you want or just something that you *think* you want.

The best way to get things in motion in what is popularly termed 'manifesting' or 'attracting' your goal is to have one intention. Do not let subconscious negativity arise. I.E. the opposite view of the intended goal, then let the intention go. It sounds simple, doesn't it? But how can we know what is going on in the subconscious? It is hidden for a reason. If everything in the subconscious were to become apparent we would not be able to function. There would be overkill. It would be like reading a book that has the pages turned every half second. Like the famous phrase uttered by Mia Wallace in Pulp Fiction, it is "Too much information." So it is hidden for a reason, and this is where questions come in. They allow the required information to bubble up from the depths. Then all we need to do is be 'still' enough to see it.

Utmost concentration *can* enliven the positive aspect of the intention leaving no room for the negative to arise. That way it is not necessary to know what is going on under the radar. If at the same time you can let the intention go and focus on something else entirely then you have it made. The trouble is if you are trying to manifest ten million pounds the sheer enormity of the amount will not *let* you release it. Even attempting to make a hundred pounds a day can seem overwhelming at times.

So how about as a test we start small. The following steps relate to attracting anything. As an experiment attempt to manifest something that you don't normally come across in everyday life. It could be a green cup or a red feather. You choose something that has no significance to you. All you have to do is want to see it.

- Know what you want.
- Write it down in every detail.
- See it as being yours and feel the emotion of actually having it.
- See it internally in every detail. Visualise or imagine.
- Know it is yours.
- Be detached from it and let it go. At the same time *know* you can have it.
- Then all you have to do is something that will bring it to you. That something is the physical work needed to actually earn the goal.

Try it without the last step and see if it turns up of its own accord. It may just be as a picture in a magazine or an advert on TV. Stay open to possibilities.

When entertaining acquisition of a particular goal, ask yourself:

Is this what I *really* want?

Am I ready to accept this?

Do I feel I deserve it?

Could I handle having this in my life?

Do I have an unconscious fear of success and/or failure?

These questions are okay in and of themselves but in order to get the answers, we have to let them sink deep within us. The answers will arise from our depths as that is where they are. Meditation can help or just keeping the questions in mind as we fall asleep. The natural dip into lower brain wave states will free them to arise spontaneously. They may come to the surface as we drift into sleep, as we dream or as we awaken. Just stay aware and ready for them. After all, it is a personal question that we are asking. Don't make the mistake of discounting this as mumbo jumbo. Your mind is way more powerful than you know. At least 90% of it is subconscious.

Imagine there is a "life's road map" that shows you exactly what you have the potential to achieve in your lifetime. It gives way markers and a list of high and low spots. What if at the top of the list it proclaimed the exact thing that you would be remembered for once you were gone, what do you think it would be? Do you think if you saw that map that you would try and do better than the prediction? Or would you sit back, open a beer and say, "Well, that's a certainty so I'm just going to wait for it to happen." What would you do if you knew you couldn't fail. What would you do if you had already reached that goal and didn't know what to do next?

Now imagine that you have fifty million pounds in the bank. You have done all the things like buy houses, cars, bikes and boats. You have taken all the holidays, given to charity and now you have free time.

What would you do with your life?

Once you know the answer to that, decide to get on with it. Even if it's only a start as it will bring you untold benefits. Ask yourself, "why aren't I doing it?"

They say "If you want to read a good book, write one." That applies to the way we live our lives just as much. If we want to have something happen, a career, a relationship, whatever, we have to *make* it happen. It won't just mooch along and happen to us, I can't emphasise this enough. The bad news, and also the good news is, we each are the only ones responsible for the quality of our life. If we find ourselves dragging our feet, we must give ourselves a slap round the head. No one else has the authority or the reason to do that for us.

Success is a journey, not a destination. Sometimes that journey will take unexpected twists and turns and dump us in a foreign land. Never let that be the end as there is always more. The universe is expanding and in order to flourish, we need to go along with that. In order to survive ,we need to grow and to evolve.

So there we are, the end of our tale is nigh. As I was pondering the note to finish on I got to thinking what this book is really all about. I didn't start out with a concrete plan as it wasn't possible to know what was going to happen. On review, I see that in essence, it is about one thing. We can take a machine and work on it. We can customise and tweak its performance and the way it works. We can then use that machine to accomplish something in our everyday world. This relates not only to metal but to our mind. We can invest in ourselves and learn continually. We can reach for higher and higher goals. This does not mean we will hit those goals, but it does mean something will inevitably happen. Something *always* happens. But are we letting it happen or are we taking the throttle of life in our hand and making it happen? It is down to a decision and each of us has the ability to make one of those.

Wishing you all that you wish for yourself, may you live a life of freedom, purpose and ease. Remember, don't be a square, Dadio.

Rod Lawless.

Other titles by Mr Lawless:

Fiction.
The Tao of Touring
Barn Find.

Non Fiction.
The Secret to Finding Love.

Reaching Your Creative Potential.

A View of The Path.

Pre-school Children's.
The Fly of The Potomi.

Power Journal's for Accomplishment, Dreams and Creativity. Available in black and white or colour for more sub-conscious power. If you keep a diary, these will give the practice more energy for change in your life.

Visit the website for more pictures, stories and video. Www.rodlawless.org

Author, Rod lawless and
Paul Anderson, builder of Svarog.